Relationship Inference with Familias and R

Relationship Inference with Familias and R
Statistical Methods in Forensic Genetics

Thore Egeland

Daniel Kling

Petter Mostad

AMSTERDAM • BOSTON • HEIDELBERG • LONDON
NEW YORK • OXFORD • PARIS • SAN DIEGO
SAN FRANCISCO • SINGAPORE • SYDNEY • TOKYO
Academic Press is an imprint of Elsevier

Academic Press is an imprint of Elsevier
125 London Wall, London, EC2Y 5AS, UK
525 B Street, Suite 1800, San Diego, CA 92101-4495, USA
225 Wyman Street,Waltham,MA 02451, USA
The Boulevard, Langford Lane, Kidlington, Oxford OX5 1GB, UK

Notices
Knowledge and best practice in this field are constantly changing. As new research and experience broaden our understanding, changes in research methods, professional practices, or medical treatment may become necessary.

Practitioners and researchers must always rely on their own experience and knowledge in evaluating and using any information, methods, compounds, or experiments described herein. In using such information or methods they should be mindful of their own safety and the safety of others, including parties for whom they have a professional responsibility.

To the fullest extent of the law, neither the Publisher nor the authors, contributors, or editors, assume any liability for any injury and/or damage to persons or property as a matter of products liability, negligence or otherwise, or from any use or operation of any methods, products, instructions, or ideas contained in the material herein.

Library of Congress Cataloging-in-Publication Data
A catalog record for this book is available from the Library of Congress

British Library Cataloguing in Publication Data
A catalogue record for this book is available from the British Library

ISBN: 978-0-12-802402-7

For information on all Academic Press publications
visit our website at http://store.elsevier.com/

Typeset by SPi Global, India
www.spi-global.com

Printed and bound in the United States

Publisher: Sara Tenney
Acquisitions Editor: Elizabeth Brown
Editorial Project Manager: Joslyn Chaiprasert-Paguio
Production Project Manager: Lisa Jones
Designer: Mark Rogers

Contents

Preface

Given DNA data and possibly additional information such as age on a number of individuals, we may ask the question: "How are these people related"? This book presents methods and freely available software to address this problem, emphasizing statistical methods and implementation. Relationship inference is crucial in many applications. Resolving paternity cases and more distant family relationships is the core application of this book. Similar methods are relevant also in medical genetics. The objective may then be to find genetic causes for disease on the basis of data from families. It is important to confirm that family relationships are correct, as erroneously assuming relationships can lead to misguided conclusions. From a technical point of view, there are similarities between the methods and software used in forensics and those used in medical genetics.

Relationship inference is not restricted to human applications. In fact, the last of four motivating examples in the first chapter is a "a paternity case for wine lovers" involving the relationship of wine grapes. Furthermore, the software presented in this book has been used in, for instance, determination of parenthood in fishes and bears. The underlying principles are then the same.

The book consists of eight chapters with exercises (except for Chapter 1) and a glossary (for nonbiologists). Chapter 1, 2, and 5 are intended to be elementary, Chapters 3 and 4 are a bit more challenging, while Chapters 6–8 are more theoretical. Chapter 2 and selected parts of Chapters 3–5 are well suited for courses for participants with a modest background in statistics and mathematics. Selected parts of the remaining chapters could be used in undergraduate and graduate courses in forensic statistics. Some new scientific results are presented, and in some cases new arguments are given for published results.

The book's companion website http://familias.name contains information on the software, tutorials, solutions to the exercises, videos, and links to a large number of courses, past and present. All software used in the book is freely available, which we consider to be an important aspect; once you have the book, you will have access to all the information and tools that are needed to do all the problems we cover. Furthermore, some of the theoretical derivations, in addition to providing a better understanding, may be used for validation purposes.

ACKNOWLEDGMENTS

A number of colleagues and friends have contributed in different ways. Magnus Dehli Vigeland has helped in many ways, and he deserves special thanks for extending his `R` package `paramlink` to cover our needs. It is a pleasure to thank Mikkel Meyer Andersen, Robert Cowell, Jiří Drábek, Guro Dørum, Maarten Kruijver, Manuel García-Magariños, Klaas Slooten, Andreas Tillmar, and Torben Tvedebrink. We are grateful for help and understanding from colleagues and students. The work of Thore Egeland leading to these results was financially supported by the European Union Seventh Framework Programme (FP7/2007-2013) under grant agreement no. 285487 (EUROFORGEN-NoE).

CHAPTER

Introduction

1

CHAPTER OUTLINE

A child inherits half its DNA from its mother and half from its father. It follows that information about the DNA of a set of persons may provide information about how they are related. The simplest and commonest example is that of paternity investigations, in which the question is whether a man is the biological father of a child. Usually, DNA tests of the mother, child, and alleged father together provide strong evidence for or against paternity. However, because of biology being variable and full of exceptions, DNA tests can never provide 100% certain conclusions in either direction (although sometimes one can get quite close). Among the thousands of paternity investigations done every year, quite a few will have somewhat ambiguous results. In such cases, statistical models and calculations can help provide reliable conclusions.

In the study of the more general question of how a set of persons are related, the strength of the evidence from DNA data may often be much weaker than in paternity cases. For example, if the question is whether two persons are cousins or unrelated, DNA test data from the two will generally not provide conclusive evidence in either direction, and statistical calculations of the strength of evidence become crucial. This is also the case when the available DNA data is limited or may contain errors, as may happen for example when some of the DNA data is based on traces from dead or missing persons.

There are a wide range of applications of relationship inference. Many types of relationships beyond paternity may be questioned and investigated for emotional, legal, medical, historical, or other reasons. The central goal may be that of identification: for instance, one may identify a dead body as a missing person by comparing DNA from the dead body with DNA from the missing person's relatives. There are also more technical uses of relationship inference: For example, in medical linkage analysis, where the goal is to reveal possible genetic causes of a disease, it is essential that relationships between the persons tested are correctly specified. In other words, information about their relationships or lack of such should be inferred from the DNA data and compared with reported information. Finally, relationship inference is also relevant for species other than humans. It has been applied to a number of animal species, and even to wine grapes [1].

This book aims to describe and discuss a statistical framework for relationship inference based on DNA data. The goal is to give the reader a comprehensive theoretical understanding of some of the most commonly used models, but also to enable her or him to perform the statistical calculations on real-life case data. Although some simple calculations can be done by hand, most are in practice done with the aid of specialized computer tools. Our own work on relationship inference [2–11] has been closely linked to developing and providing free software. The program `pater` was released in 1995. In 2000 the name of the program changed to `Familias`, and it is currently one of the most widely used tools for statistical calculations in DNA laboratories [12]. Further Windows programs (`FamLink` and `FamLinkX`) have been developed more recently. There is also an `R` package[1] called `Familias`, implementing the same core functionality as the Windows program. Theory and computational methods will primarily be illustrated and practiced with these programs. However, we will also use a number of additional `R` packages that implement various useful functions, such as `disclap`, `disclapmix`, `DNAprofiles`, `DNAtools`, `identity`, `kinship2`, and `paramlink`.

Apart from relationship inference, DNA tests of the type mentioned above are often used for identification purposes—for example, in criminal investigations. Again, computation of the strength of the evidence is important. Many issues are similar in the two applications, although issues concerning missing or degraded DNA, or mixtures of DNA from several persons come to the fore in criminal investigations. Forensic genetics encompasses all applications of DNA tests to questions such as identification and relationship inference. A number of books (e.g., [13–16]) deal with this perspective. In addition, forensic statistics more generally is addressed in [17–19]. There is also another line of literature, not considered in this book, where the framework of Bayesian networks is successfully used to deal with forensic problems; see [9, 20, 21].

In this book, we focus more narrowly on the problem of relationship inference based on DNA data. This gives us the opportunity to describe and discuss some topics that may otherwise be hidden in the specialized literature. Also, some well-known theory may be phrased in new ways.

1.1 USING THIS BOOK

Our intended audience includes several groups. Firstly, we would like to provide case workers in forensic laboratories with a central reference and tool for training and study. Secondly, we hope scientists involved in teaching or research in this area

[1] http://www.r-project.org/.

will find our theoretical material and our exercises interesting and useful. In some research, solving questions about disputed relationships may be a secondary problem, and researchers may then find the current text useful as an introduction and reference. We also hope statisticians with no particular background in forensic genetics will find the material interesting and readable as an example of applied statistics.

The potentially diverse readership means that various groups may put different emphasis on different parts of the book. Generally, we do not require more than a rudimentary background in statistics. Understanding simple discrete probability calculations will suffice for the study of most parts of Chapters 1, 2, 3, and 5. Exercises or material that may require some additional statistical background are marked with a star, and in a few cases with two stars to indicate even more challenging material. The remaining chapters assume knowledge of some additional statistical concepts, although readers who do not understand all the mathematical details will hopefully also find these chapters useful.

The main text will assume knowledge of a number of biological and technological concepts underpinning DNA testing. As most readers are likely to be familiar with these, we have chosen not to discuss them at any length; however, we have included a glossary which aims to provide the information necessary to read the book even with no biological or technological background beyond a minimal general knowledge of DNA.

We have included a large number of exercises, to the benefit of those who prefer to learn by doing exercises. The companion online resources for the book can be found via the website http://familias.name. You may find there input files for exercises, suggested solutions, and tutorial videos for the various programs we use. The programs themselves may be downloaded (freely) from their corresponding websites: http://familias.no for `Familias` and http://famlink.se for `FamLink`, and `FamLinkX`. The `R` packages can be downloaded from the Comprehensive `R` Archive Network; see http://r-project.org. The Windows programs are intended to be easy to use for anybody, whereas use of `R` packages requires some familiarity with `R`. Chapters 1–4 do not use `R`, but starting from Chapter 5, `R` is the main tool illustrating theory and computations. We do not include an `R` tutorial as many excellent tutorials for people of different backgrounds are available online. Although the theory in Chapters 5–8 may be read without knowing `R`, we encourage readers who do not yet know this program to become familiar with it. In many examples, we illustrate how easily `R` can be used to build new ideas and extensions on top of old methods, making it an invaluable tool for a researcher.

Chapter 2 first explains the basic methods, starting with a standard paternity case. The examples and most exercises use the Windows version of `Familias`; a tutorial is available at http://familias.name. The chapters that follow provide extensions in various directions. Searching for relationships in a greater context, such as disaster victim identification and familial searching are discussed in Chapter 3. Chapter 4 considers dependent markers, where examples and exercises are based on the programs `FamLink` and `FamLinkX`, and it is demonstrated how relevant problems can be solved. For instance, with use of X-chromosomal markers, it becomes possible to distinguish maternal half-sisters from paternal ones.

Chapter 5 introduces R functions implementing many of the computations from previous chapters, while Chapters 6–8 present the theory in a more general framework. This allows for extensions, and some previous simplifying assumptions can be removed. For instance, the first four chapters assume allele frequencies to be known exactly. More generally, uncertainty in parameters can be accommodated, as explained in Chapter 7. Forensic testing problems can be seen as more general decision problems as explained in Chapter 8.

1.2 WARM-UP EXAMPLES

Four examples corresponding to Figures 1.1–1.4 are presented briefly, with a detailed discussion being deferred to later sections. The purpose is to delineate more precisely the problems we seek to provide solutions for. Words and concepts that may be unknown to some readers are defined and discussed in Chapter 2.

Example 1.1 Paternity (introductory example). Figure 1.1 shows a standard paternity case discussed further in Section 2.2. Data for one genetic marker is given. In this case, the genotypes are consistent with the alleged father being the biological father as shown in the left panel since the alleged father and the child share the allele denoted A. Typically data will be available for several markers, say at least 16. It may happen that all markers but one are consistent with paternity, while the last indicates otherwise. A standard calculation will give a likelihood ratio of 0, resulting in an exclusion. However, mutations cannot be ignored and should be accounted for. This will dramatically change the result and the conclusion regarding paternity.

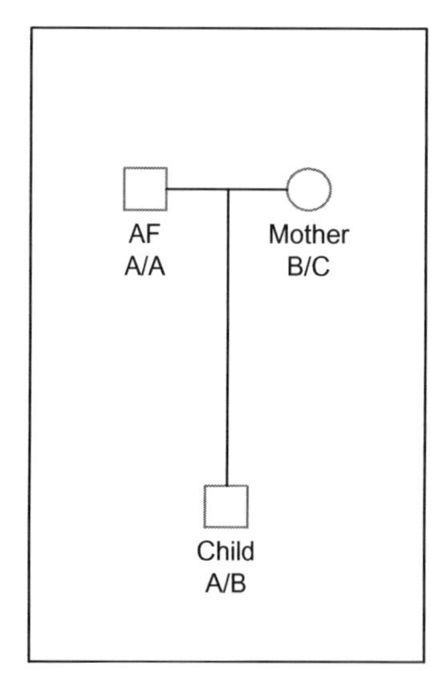

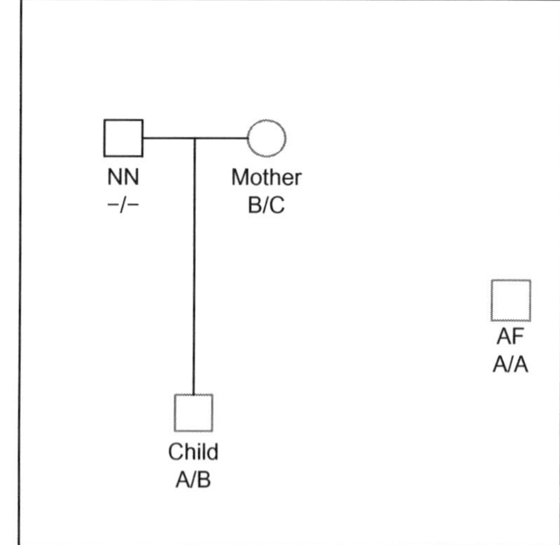

FIGURE 1.1

A standard paternity case. The left panel corresponds to hypothesis H_1, the alleged father (AF) being the father. In the right panel, the alleged father is unrelated to the child (hypothesis H_2).

Example 1.2 Missing person (dropout?). Figure 1.2 displays a case with a missing person: A body (denoted 4 in the figure) has been found. There are two hypotheses corresponding to the two panels in the figure. The body has been in a car underwater for 20 years, resulting in a suboptimal DNA profile for 4 as indicated by the genotype 1/−. This means that only one allele, named 1, is observed, while the other allele may have dropped out. To determine whether the missing person has been found, corresponding to the pedigree to the left, advanced models and software are needed. Sometimes additional complications must be accounted for: an allele may fail to amplify, there may be deviations from *Hardy-Weinberg* equilibrium, and there may be uncertainty in parameters such as allele frequencies.

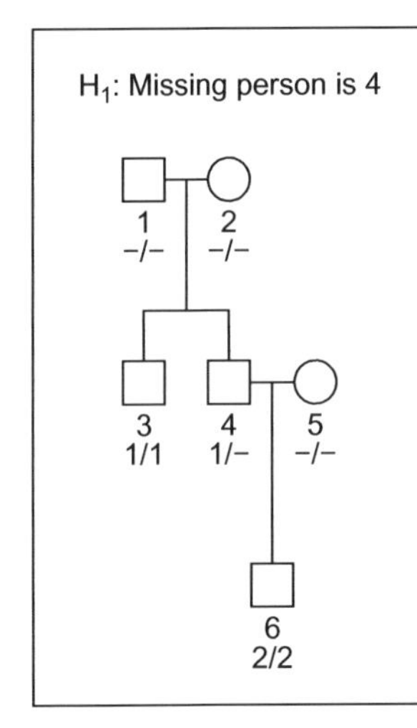

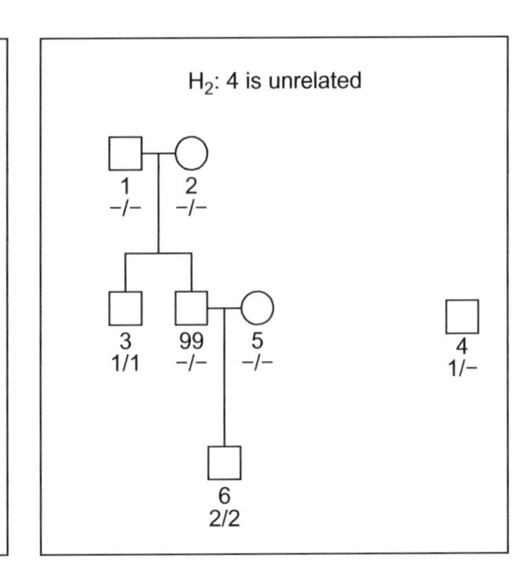

FIGURE 1.2

A case of a missing person. Is individual 4 the brother of 3 and the father of 6 (left panel) or an unrelated person (right panel)?

Example 1.3 Disaster victim identification. In Figure 1.3, a disaster victim identification problem is depicted. There are three deceased individuals and two families F1 and F2. The data points to V1 being missing from F2, while V2 belongs to F1; individual V3 appears not to belong to either F1 or F2. Disaster victim identification problems are closely related to relationships problems, and are therefore conveniently implemented in the same software. However, a large number of hypotheses are sometimes compared, and this leads to methodological and computational challenges which are addressed in Chapter 3.

The examples so far have considered data only for one marker. Calculations can easily be extended to several markers that are assumed to be independent. However, if independence cannot be assumed, matters are more complicated, as discussed in Chapter 4.

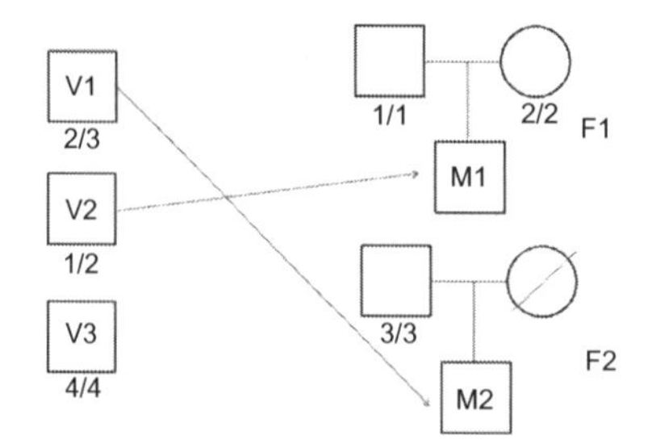

FIGURE 1.3

A matching procedure in a disaster victim identification operation. V1, V2, and V3 denote victims, while M1 (in F1) and M2 (in F2) denote missing persons.

Example 1.4 A paternity case for wine lovers. The three examples above deal with human applications. Similar methods and software can be used for problems involving animals or plants. Figure 1.4 describes a case referred to as "a paternity case for wine lovers" in [22], and deals with the origins of the classic European wine grape *Vitis vinifera*. Again, several hypotheses are considered; some may be likelier than others on the basis of non-DNA data, and this can be accounted for by introducing a *prior* distribution. The prior can be combined with the likelihood of the data to obtain the *posterior* distribution. The most probable pedigree is found, and this is an alternative to reporting the likelihood ratio. Further background and details are given in Section 2.12.2.

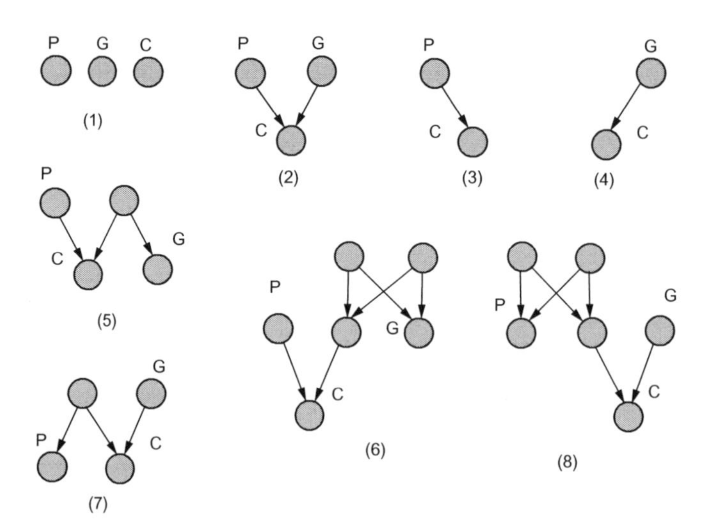

FIGURE 1.4

A paternity case for wine grapes showing eight alternative pedigrees for the relationship of Chardonnay (C) with Pinot (P) and Gouais blanc (G).

1.3 STATISTICS AND THE LAW

Our topic is part of forensic statistics, which concerns the intersection of the areas of statistics and law, and so it may be appropriate to discuss briefly the relationship between these two fields. We first note that "statistical methods," appearing in the title of this book, belong to (applied) mathematics. Statistical methods rely on probability theory and address "how conclusions are drawn from data" [23]. Tribe [24] writes in the widely cited and much discussed paper "Trial by mathematics: precision and ritual in the legal process":

> *I am, of course, aware that* all *factual evidence is ultimately "statistical" and all legal proof ultimately "probabilistic", in the epistemological sense that no conclusion can ever be drawn from empirical data without some step of inductive inference—even if only an inference that things are usually what they are perceived to be.*

The applications that we have in mind for the methods and implementation presented in this book are not limited to trials or legal contexts. For instance, "relationship inference" may be performed by persons reconstructing their family pedigree for purely personal reasons. The methods used in such private settings may well coincide with those presented in a court of law. However, for this section legal applications are central, and we discuss some principles that may be relevant for those doing work with potential legal applications. These principles are not limited to analyses based on genetic data. However, forensic genetics has been a driving force also when it comes to more principle issues as noted in [25]: "The traditional forensic sciences need look no further than their newest sister discipline, DNA typing, for guidance on how to put the science into forensic identification science."

1.3.1 CONTEXT

The legal systems differ between countries, and it is common to distinguish between the adversarial legal system of the US, the UK, and other English-speaking countries and the inquisitorial system common in large parts of mainland Europe. Typically, each party will be represented by its own scientific expert in the adversarial system, whereas by default there is only one expert in the inquisitorial system. While these different traditions may have wide-ranging implications for court procedures, the presentation in this book is not influenced by this distinction. Statements such as "the statistician must respect the concept that representation of the client's interest is in the hands of the attorney" [23] may be considered appropriate and relevant by some in an adversarial context. In contrast, the guiding principle of the inquisitorial system is to be unbiased and independent.

1.3.2 TERMINOLOGY

The formulation of two competing hypothesis is a key ingredient and common starting point for statistical analyses in forensic applications. In Figure 1.1 these hypotheses are denoted H_1 and H_2. H_P and H_D are common alternatives with "P" and "D" referring to the prosecution and defense hypotheses, respectively. This terminology is even used when there is no obvious reference to different parties in a court case. Only rarely will the parties representing the prosecution and defense be consulted before the hypotheses are formulated. Rather, the hypotheses are needed to get the calculations started. We prefer the more neutral versions H_1 and H_2.

1.3.3 PRINCIPLES

The following principles for evaluation of evidence were formulated in [16]:

1. To evaluate the uncertainty of any given proposition it is necessary to consider at least one alternative proposition.
2. Scientific interpretation is based on questions of the following kind: What is the probability of the evidence given the proposition?
3. Scientific evidence is conditioned not only by the competing propositions, but also by the framework of circumstances within which they are to be evaluated.

The first principle is nicely illustrated by a Norwegian Supreme Court case. The question was whether the use of the contraceptive pill had caused the death of a woman. In the ruling it was argued that the probability that the pill had caused the death was very small. Mainly for this reason the company producing the pill was acquitted. However, this statement carries little evidentiary value unless other possible explanations for the death of the woman are considered: all other possible explanations could be even less likely.[2] There are different published versions of the above principles. For instance, in [26] which precedes [16], principles 1 and 2 resemble those above, but principle 3 reads as follows:

> *The strength of the evidence in relation to one of the explanations is the probability of the evidence given that explanation, divided by the probability of the evidence given the alternative explanation.*

This version explicitly declares that the likelihood ratio, which plays an important role throughout this book, should be used. It is essential that this version and also principle 2 above state that the expert should report on the evidence given the hypotheses.

[2]The verdict discussed is published in the periodical published by the Norwegian Bar Association: *Retstidende*, 1974, p. 1160 (available only in Norwegian).

1.3.4 FALLACIES

Sometimes the conditionals are transposed, resulting in a statement of guilt given the evidence, or more commonly, the statement from the expert is misinterpreted—for instance, by the prosecutor. This fallacy is sufficiently common to have earned a name: "the prosecutor's fallacy." There is also a defense attorney's fallacy and a typical version is as follows: "The frequency of the defendant's DNA profile is 1 in a million. There are 10 million males that could have left the stain. In other words, we can expect ten people to match. The probability that the stain comes from the defendant is thus 1 in 10." The problem with this statement is that it ignores the context, principle 3 above. There is a reason why the defendant has been taken to court. If he had rather been found after a database search, evaluating the strength of the evidence becomes more complicated, as explained in [27, 28]. Gill [29] explores the fallacies mentioned above in greater detail, and also presents and exemplifies other fallacies. The latter example can easily be formulated in a paternity context, more fitting for the applications in this book: if a sufficiently large male population is considered, several men could fit as a father.

CHAPTER

Basics

2

CHAPTER OUTLINE

This chapter describes the data and the basic methods. More advanced methods are presented in later chapters. The core example is a paternity case. Competing hypotheses are formulated—for instance, that a man (alleged father) is the father of a child versus some man unrelated to the alleged father is the father. The statistical evidence is normally summarized by the likelihood ratio (LR). For instance, $\text{LR} = 100,000$ implies that the data is 100,000 likelier if the man is the father compared with the alternative. The LR may be converted to the probability of paternity, the posterior probability, which relies on the prior probability of paternity.

In some cases, there is a need to go beyond the standard kits of autosomal markers. In brief sections we discuss the X chromosome and lineage markers (Y chromosome and *mitochondrial DNA*; mtDNA).

The core methods in this chapter are presented in textbooks such as [13–16]. Several factors such as mutation, theta correction, silent alleles, and dropout may complicate calculations, and the required extensions are discussed and references are provided.

2.1 FORENSIC MARKERS

Below, we summarize the basic facts we will need about forensic markers. *Note*: Some biological and technological terms that are mentioned only briefly are discussed further or defined in the glossary; this may then be indicated with the use of *italics*.

When measuring DNA in order to infer relationships, one investigates only a very small part of the total DNA sequence. Apart from technological and economic issues, the main reason is that in most cases only very small parts are needed to reach a conclusion. The parts that are investigated are called forensic *markers*. Any two humans of the same sex have DNA sequences that are more than 99% identical.[1] A forensic marker is characterized as a location along those sequences where differences may occur, so it may also be called a *locus*. The usefulness of a forensic marker in our context depends on its *polymorphism*—that is, how much variability there is at this location. A particular variant of a marker is called an *allele*. The more alleles there are for a marker, the likelier it is that unrelated persons will have different DNA at the location, and thus matching DNA more strongly indicates a relationship.

Knowledge of the connection between the DNA sequence and the human *phenotype*—for example, disease status—is rapidly developing. Traditionally, forensic markers have been chosen so that marker variability has no known connection with variations in the phenotype. Obtaining phenotypic or medical information about someone during a forensic investigation is in most cases an unwanted ethical or legal complication. There are cases, for example, connected to identification where such information may be welcomed, and research continues into developing such markers;

[1] http://en.wikipedia.org/wiki/Human_genetic_variation.

see [30]. However, in this book we will stick to markers with no known phenotypic interpretation.

The evolutionary process that has created the human DNA sequence and the variability in it is of course continuing. Changes in DNA sequences are called *mutations*. They can be divided into *somatic* mutations, which affect only the individual in which they occur, and *germ line* mutations, which are passed on to the next generation; we will be concerned with only the last type. Mutations can happen anywhere in the *genome*, but will happen with different probability; the probability for a change in the DNA from one generation to the next at a locus is called the *mutation* rate of the locus. Highly polymorphic markers tend to be polymorphic because they have a high mutation rate. So when such markers are used to infer relationships, it often becomes necessary to take into account the possibility of a mutation within the case data under consideration.

In addition to being polymorphic, a forensic marker needs to have a technological measurement process with which its alleles can be determined reliably at a reasonable cost. We will not be concerned with the details of these processes except when they might contain measurement errors. This happens most often when the data is based on degraded or unusually small amounts of DNA, and is commoner in criminal cases or identification cases than in cases where relationship inference is the main focus. We will return to this subject in Sections 2.7 and 2.9.3.

Finally, to use data from forensic markers to check for possible relationships, we need to know what data to expect for unrelated persons. Specifically, one needs databases tallying the alleles of large numbers of unrelated persons from various populations, from which one can compute population frequencies for the alleles. Thus, standardized sets of forensic markers are useful so that laboratories in various countries and areas can compare and pool their data. Such standardization is also driven by the existence of commercial *typing kits*, where the alleles of 10-30 markers can be determined in a single procedure. In the choice of a forensic marker, it is often valuable that its population frequencies do not vary too much between populations. If they do, inference from these markers will depend on assumptions about which population the persons in a case come from.

Our two main examples of forensic markers will be *single nucleotide polymorphism* (SNPs) and *single tandem repeats* (STRs) (see the glossary definitions). Large number of SNP markers can be measured simultaneously by either *microarray technology* or *next-generation sequencing*. Their usefulness rests on their large numbers and low mutation rates. However, there are in most cases only two observed alleles for each marker. Thus, it is far commoner to use STR markers in forensic investigations, where the information from each marker is greater. There are standardized sets of such markers containing from six to more than 20 markers, where each marker has from 6 to 40 alleles; for some markers such as SE33 there may be even more alleles.

Example 2.1 Allele frequencies. In Example 2.2, we use data from the STR markers D3S1358 and TPOX. Figure 2.1 shows the observed alleles, with their proportions of observation, based on a Norwegian frequency database. In particular,

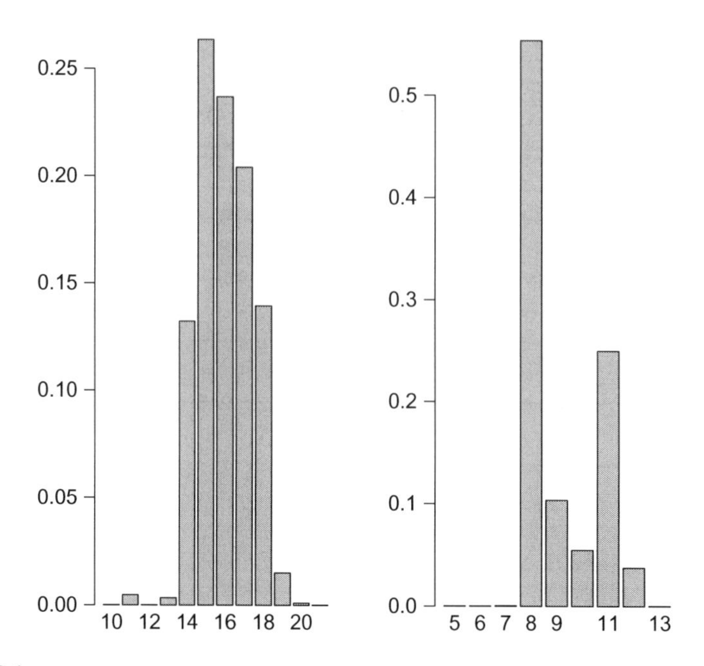

FIGURE 2.1

Alleles and allele frequencies for STR markers D3S1358 (left panel) and TPOX.

the population frequencies of alleles 17 and 18 in marker D3S1358 are 0.2040 and 0.1394, respectively, and the allele frequency of allele 8 in marker TPOX is 0.5539. The complete database is available as the dataset `NorwegianFrequencies` in the `R` version of `Familias` as explained in Chapter 5.

The utility of forensic markers for relationship inference depends on how they are inherited. Most forensic markers are located on *autosomal chromosomes*, so each person inherits two copies of a locus, one from the mother and one from the father. The inherited alleles are called the *maternal allele* and the *paternal allele*, respectively. For any autosomal marker, there is a 50% chance for each of the two alleles to be passed on to a child. For markers located close to each other on the same *chromosome*, there may be a dependency (*linkage*) concerning which alleles are passed on to a child. The two markers are then said to be linked, and we will discuss this further in Section 4.1.1. In some cases of relationship inference there is a need for a large number of markers, and it may be difficult to avoid the use of linked markers. The special properties of linked markers may also be useful in some cases. But traditional standard marker sets are chosen so that markers are (more or less) unlinked, and in this chapter and Chapter 3 we will assume all markers are unlinked.

The question of dependency between alleles at different markers is, however, more general. The process of evolution, and how alleles are spreading in different populations, is very complex. If we observe in a person an allele which varies in frequency between different populations, it increases the likelihood that the person is a member of a certain population relative to other populations. This, in turn, increases the probability that she or he carries alleles that are more frequent in those populations, even at markers other than that of the first observed allele. General dependency between alleles at different markers is called *linkage disequilibrium*, and will be discussed further in Section 4.2. In this chapter and Chapter 3, we will make the simplifying assumption that markers are completely independent of each other; we assume there is linkage equilibrium.

In addition to autosomal markers, there are also forensic markers on the sex chromosomes X and Y, and within the *mitochondrial genome*. We will return to such markers in Section 2.9.

Throughout this book we do not consider anomalies such as *trisomies*; see Section 1.2.1.1 in Buckleton and Gill [14].

2.2 PROBABILITIES OF GENOTYPES

Before we can do probability calculations for relationship inference, we need to start with a more basic question: What is the probability of observing a particular *genotype* at a certain locus—for example, what is the probability of observing the genotype 17/18 at the locus D3S1358? (In other words, what is the probability that a person has alleles 17 and 18 at the locus, with one being the maternal allele and one being the paternal allele, but without knowing which is which?)

To calculate this, we assume we know the probability of observing each of these alleles in the population, and we assume this probability is equal to the population frequency (see Section 6.1.1 for a more thorough discussion). We also assume that the probability of observing one of the alleles is independent of that of observing the other. This may be called the assumption of *Hardy-Weinberg equilibrium* (HWE). Let us denote the population frequencies of two alleles a and b by p_a and p_b, respectively. Then, under the assumption of HWE, the probability of observing an individual with genotype a/a (an individual *homozygous* in allele a) is p_a^2, and the probability of observing an individual with genotype a/b (a *heterozygous* individual with alleles a and b) is $2p_ap_b$. The reason for the factor 2 is that the genotype a/b is compatible with both allele a being the paternal allele and allele b being the maternal allele, and the opposite alternative. Reiterating the formulas for reference, we have

$$\Pr(G = a/b) = 2p_ap_b, \tag{2.1}$$

$$\Pr(G = a/a) = p_a^2. \tag{2.2}$$

Example 2.2 Genotype probabilities. We illustrate this with some computations connected to a real paternity case, which we will return to several times. What is the probability of observing an individual with genotype $17/18$ at marker D3S1358, and genotype $8/8$ at marker TPOX?

According to Example 2.1, the allele frequencies of alleles 17 and 18 in marker D3S1358 are $p_{17} = 0.2040$ and $p_{18} = 0.1394$, respectively, and the allele frequency of allele 8 in marker TPOX is $p_8 = 0.5539$. The equations above give

$$\Pr(G = 17/18) = 2p_{17}p_{18} = 2 \times 0.2040 \times 0.1394 = 0.0569,$$
$$\Pr(G = 8/8) = p_8^2 = 0.5539^2 = 0.3068.$$

Assuming there is linkage equilibrium, we get that the probability of observing both genotypes in one person is

$$0.0569 \times 0.3068 = 0.0175.$$

2.3 LIKELIHOODS AND LRs

The likelihood is the probability of the data conditional on the hypothesis and the parameters. The numerator and denominator of the LR are both likelihoods, and programs such as `Familias` compute and list these likelihoods using for example the *Elston-Stewart algorithm* [31]. For linked markers, the Lander-Green algorithm described in Section 4.1.3 is central. The calculations simplify greatly for noninbred cases involving only two genotyped individuals, as explained in Section 2.3.3.

2.3.1 STANDARD HYPOTHESES

The statistical treatment of paternity cases normally starts with the formulation of two competing hypotheses:

H_1: The alleged father is the biological father of the child.
H_2: A random man is the biological father.

Hypothesis H_1 with genotypes included is shown in Figure 2.2. We will refer to cases specified by the above hypotheses as a standard duo case. Only the child and alleged father will be genotyped. We have used "biological father" above. Occasionally, alternatives like "real father" or "true father" are seen. By "random man" we imply that the genotypes of the man is randomly sampled from the relevant database. Sometimes, "unrelated" is used rather than "random".

For autosomal markers, the genders of the persons involved does not, for practical purposes, matter. However, females have two X chromosomes, while males only have one, and therefore the inheritance patterns differ and gender information is needed and used as elaborated on in Section 4.1.4. Frequently, the undisputed mother is genotyped, in which case we use the term *trio case*. The precise formulation of the hypothesis is crucial. If the alternative hypothesis H_2 includes relatives of the alleged father, such as a brother, the assessment of the evidence would change dramatically. For the data at hand, the alleged father and the child share alleles for both markers. A relative of the alleged father would be likelier than an unrelated man to also share

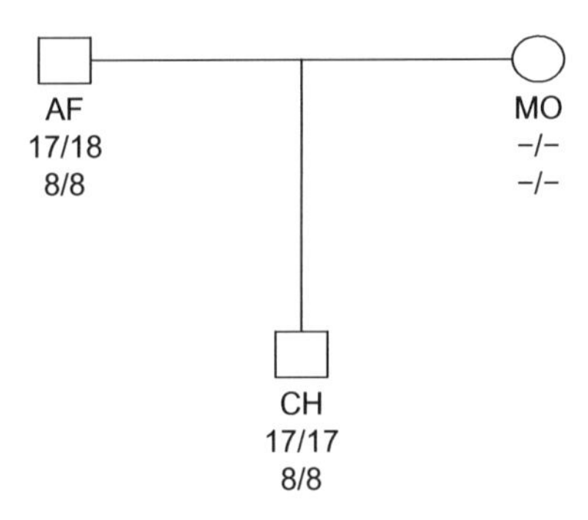

FIGURE 2.2

Pedigree for duo case. AF, alleged father; CH, child; MO, mother.

alleles with the child, and therefore the evidence in favor of paternity would be weaker if relatives of the alleged father are possible alternative fathers.

The formulation and testing of hypotheses in forensic genetics typically differ from that in other areas such as medical statistics. A few comments are therefore in order. First, we note that forensic hypotheses, as the ones above, are formulated verbally. In most other areas, the hypotheses are written up with the parameters of a statistical model. Furthermore, one of the hypotheses is normally referred to as the null hypothesis. For instance, the problem of testing if the expected value μ of a normal distribution equals 0 is normally formulated $H_0 : \mu = 0$. A p-value is typically calculated, and the null hypothesis is rejected if this p value is below some prescribed threshold, typically 0.05. This implies an asymmetry between the hypotheses. It is considered most important to avoid falsely rejecting the null hypothesis, and it is this error that is controlled by requiring a small p-value. Forensic problems differ in being symmetric: it is normally equally important to avoid rejection of either hypothesis, although we will return to a discussion of this at the beginning of Chapter 8.

The principal requirement for relationship testing is not to prove a relationship beyond reasonable doubt, but rather to determine the likelier hypothesis. This being said, there will normally not remain any reasonable doubt. For the above reason, we deliberately avoid referring to a null hypothesis; there is none. Rather than calculating p-values, we find an LR, as explained next.

2.3.2 THE LR

A simple approach to the calculations is presented first, followed by a more careful general derivation included to highlight the assumptions. The marker D3S1358 is considered to begin with. Assume H_1 (see Figure 2.2) is true. The father then passes on either allele 17 or allele 18 to the child, each with probability 0.5 according to Mendel's law of independent assortment. In addition, the child must receive a copy of allele 17 from the mother, and this happens independently with probability p_{17}. The frequency of the allele in the population is used as there is no information on the genotype of the mother. Therefore, the genotype probability of the child equals $0.5 \times p_{17}$. For the alternative hypothesis, there is no information on parent genotypes, and the probability is p_{17}^2. The LR is therefore

$$\mathrm{LR}_1 = \frac{0.5 \times p_{17}}{p_{17}^2} = \frac{1}{2p_{17}} = \frac{1}{2 \times 0.20405} = 2.4504. \tag{2.3}$$

For the second marker, the probability that the child is 8/8 given that the father is 8/8 is p_8, and

$$\mathrm{LR}_2 = \frac{p_8}{p_8^2} = \frac{1}{p_8} = \frac{1}{0.55391} = 1.8053.$$

The combined LR is obtained by multiplication as the markers are independent, and therefore $\mathrm{LR} = 2.4504 \times 1.8053 = 4.42$. The interpretation is that the data is 4.42 times likelier if the alleged father is the father compared with the unrelated alternative.

Next the more precise derivation follows. Recall that the LR is the probability of the data given H_1 divided by the probability of the data given H_2. This is expressed more formally as

$$\begin{aligned} \mathrm{LR} &= \frac{\Pr(\text{data} \mid H_1)}{\Pr(\text{data} \mid H_2)} = \frac{\Pr(G_{\mathrm{AF}}, G_{\mathrm{CH}} \mid H_1)}{\Pr(G_{\mathrm{AF}}, G_{\mathrm{CH}} \mid H_2)} \\ &= \frac{\Pr(G_{\mathrm{CH}} \mid G_{\mathrm{AF}}, H_1)}{\Pr(G_{\mathrm{CH}} \mid G_{\mathrm{AF}}, H_2)} \times \frac{\Pr(G_{\mathrm{AF}} \mid H_1)}{\Pr(G_{\mathrm{AF}} \mid H_2)}. \end{aligned} \tag{2.4}$$

We will discuss the different parts of the above expression in turn since this will make the assumptions clear. Generally, the probability of the genotype G_{AF} does not depend on the hypotheses. Therefore,

$$\Pr(G_{\mathrm{AF}} \mid H_1) = \Pr(G_{\mathrm{AF}} \mid H_2) = \Pr(G_{\mathrm{AF}})$$

and $\Pr(G_{\mathrm{AF}} \mid H_1)/\Pr(G_{\mathrm{AF}} \mid H_2) = 1$. This is normally assumed without mentioning it, although it is possible to imagine scenarios where it fails. For instance, the ratio could differ from 1 if the genetic marker contained information on infertility. Recall, however, that forensic markers are supposed not to carry phenotypic information.

Assuming that H_2 implies that the probabilities of observing the two genotypes G_{AF} and G_{CH} are independent, we get $\Pr(G_{\mathrm{CH}} \mid G_{\mathrm{AF}}, H_2) = \Pr(G_{\mathrm{CH}})$, and we arrive at

$$\mathrm{LR} = \frac{\Pr(G_{\mathrm{CH}} \mid G_{\mathrm{AF}}, H_1)}{\Pr(G_{\mathrm{CH}})}. \tag{2.5}$$

2.3.3 IDENTICAL BY DESCENT AND PAIRWISE RELATIONSHIPS

The paternity problem discussed in the previous section can be considered as a special case of a pedigree problem. The questioned family relationship could be more distant than a father-child relationship. The possibility of mutations, or incest, would also complicate the problem. Normally, computer programs are needed to perform the calculations. There is, however, an important class of problems, those involving two individuals, for which simple calculations are valid. The calculations rely heavily on the concept of *identical by descent* (IBD). An allele in one individual is IBD to an allele in another individual if it derives from the same ancestral allele within the specified pedigree. Figure 2.3 illustrates the IBD concept for brothers.

The probability that the brothers inherit different alleles from a parent is 0.5. If the parents are unrelated, the probability that they inherit different copies from both parents is $0.5 \times 0.5 = 0.25$. Similarly, the probability of no IBD sharing is 0.25. Finally, the brothers share one allele IBD with probability $1 - 0.25 - 0.25 = 0.5$. Alternatively, these IBD probabilities can be obtained by letting I denote the number of alleles shared IBD and noting that I is binomially distributed for brothers with parameters $p = 0.5$ and $n = 2$. It then follows directly that $\Pr(I = 0) = \Pr(I = 2) = 0.25$, while $\Pr(I = 1) = 0.5$. *Identical by state* (IBS) refers to allele sharing, not necessarily by descent. It may happen that parents have identical copies of alleles, and therefore the probability that the brothers share two alleles IBS exceeds 0.25.

The pedigree information for non-inbred pairwise relationships can be summarized by the IBD probabilities. This is clear from

$$\begin{aligned}\Pr(\text{data} \mid H) = {} & \Pr(I = 0)\Pr(\text{data} \mid I = 0) \\ & + \Pr(I = 1)\Pr(\text{data} \mid I = 1) \\ & + \Pr(I = 2)\Pr(\text{data} \mid I = 2),\end{aligned} \tag{2.6}$$

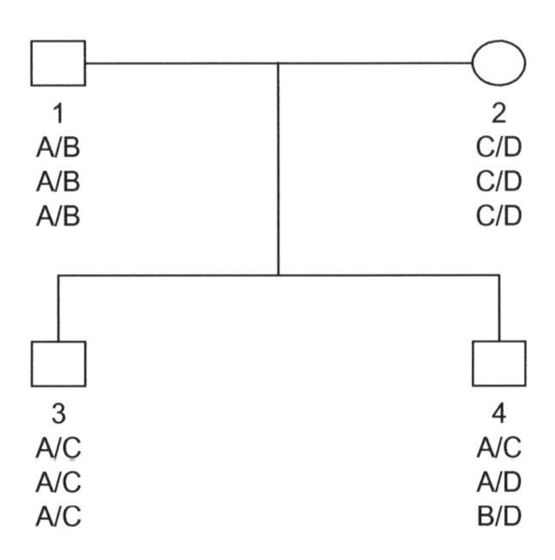

FIGURE 2.3

The brothers share two, one, and no alleles IBD, respectively, for the three markers.

Table 2.1 Probabilities for Pairs of Genotypes as a Function of the Number of Alleles Shared IBD, Indicated by I

Genotype X	$I=0$	$I=1$	$I=2$
(A/A,A/A)	p_A^4	p_A^3	p_A^2
(A/A,A/B)	$2p_A^3p_B$	$p_A^2p_B$	0
(A/A,B/B)	$p_A^2p_B^2$	0	0
(A/A,B/C)	$2p_A^2p_Bp_C$	0	0
(A/B,A/B)	$4p_A^2p_B^2$	$p_Ap_B(p_A+p_B)$	$2p_Ap_B$
(A/B,A/C)	$4p_A^2p_Bp_C$	$p_Ap_Bp_C$	0
(A/B,C/D)	$4p_Ap_Bp_Cp_D$	0	0

as the pedigree enters only via the IBD probabilities. Below we assume HWE, and then the terms $\Pr(\text{data} \mid I=i),\ i=0,1,2$ are as in Table 2.1. Several entries in the table are intuitive. For instance, the upper left entry must be p_A^4 when there is no allele sharing. Furthermore, for the last line, the alleles differ, and the probabilities must be 0 unless there is no IBD sharing. We illustrate the use of Equation 2.6 in a specific case: With the genotypes of the parents in Figure 2.3 assumed to be unknown and the first marker—that is, both brothers are A/C—consider the following hypotheses:

H_1: The individuals are brothers.
H_2: The individuals are unrelated.

On the basis of Equation 2.6 and Table 2.1, we find

$$\Pr(\text{data} \mid H_1) = \frac{1}{4}4p_A^2p_C^2 + \frac{1}{2}p_Ap_C(p_A+p_C) + \frac{1}{4}2p_Ap_C$$

$$= p_Ap_C\left(p_Ap_C + \frac{1}{2}(p_A+p_C) + \frac{1}{2}\right).$$

Since $\Pr(\text{data} \mid H_2) = 4p_A^2p_C^2$, the LR in favor of the persons being brothers becomes

$$\text{LR} = \frac{p_Ap_C + \frac{1}{2}(p_A+p_C) + \frac{1}{2}}{4p_Ap_C}.$$

If $p_A = 0.1$ and $p_C = 0.2$, then LR $= 8.375$, or approximately 8, and the interpretation is that the data is eight times likelier if the individuals are brothers compared with them being unrelated. Obviously, a much larger LR is expected in a real case with, say, 10 markers if the asserted relationship is true.

Example 2.3 Duo case (LR and IBD). Returning to the markers D3S1358 and TPOX in Example 2.1, we now let H_1 specify that the alleged father and the biological father are brothers and keep the alternative specifying them to be unrelated. Equation 2.6 combines with the equations in Table 2.1 and the genotype data in Figure 2.2 to give

$$\mathrm{LR}_1 = \frac{\frac{1}{4}2p_{17}p_{18}p_{17}^2 + \frac{1}{2}p_{17}^2p_{18}}{p_{17}^2 2p_{17}p_{18}} = \frac{p_{17}+1}{4p_{17}} = \frac{0.2040+1}{4\times 0.2040} = 1.475, \tag{2.7}$$

$$\mathrm{LR}_2 = \frac{\frac{1}{4}p_8^2p_8^2 + \frac{1}{2}p_8^3 + \frac{1}{4}p_8^2}{p_8^2p_8^2} = \frac{p_8^2+2p_8+1}{4p_8^2}$$

$$= \frac{0.5539^2 + 2\times 0.5539 + 1}{4\times 0.5539^2} = 1.968.$$

The combined LR is $1.475 \times 1.968 = 2.903$.

2.3.4 PROBABILITY OF PATERNITY: *W*

So far the evidence has been summarized by the recommended LR [32]. It is, however, possible to report the probabilities of the hypotheses given the genetic evidence. This requires prior probabilities $\Pr(H_1)$ and $\Pr(H_2)$ to be specified. Then, Bayes's theorem, which will return to repeatedly in different contexts, converts the LR to a posterior probability:

$$\Pr(H_1 \mid \text{data}) = \frac{\Pr(\text{data} \mid H_1)\Pr(H_1)}{\Pr(\text{data} \mid H_1)\Pr(H_1) + \Pr(\text{data} \mid H_2)\Pr(H_2)}. \tag{2.8}$$

In most applications a flat prior—that is, $\Pr(H_1) = \Pr(H_2) = 0.5$—is used, reflecting that the hypotheses are equally likely before data has been obtained. Bayes's theorem then simplifies to

$$\Pr(H_1 \mid \text{data}) = \frac{\Pr(\text{data} \mid H_1)}{\Pr(\text{data} \mid H_1) + \Pr(\text{data} \mid H_2)} = \frac{\mathrm{LR}}{\mathrm{LR}+1}. \tag{2.9}$$

In the forensic literature, the above probability is called the Essen-Möller index *W*, an abbreviation of the German "Warscheinlichkeit" introduced in 1938 by the Swede Essen-Möller [33, 34]. With use of this formula, $\mathrm{LR} = 4.42$ calculated previously is converted to a probability of $W = 4.42/(4.42 + 1) = 0.82$.

Probabilities such as the Essen-Möller index are constrained to the interval from 0 to 1 (or from 0% to 100%) and are easier to interpret than LRs, but this comes with a price: the need to specify prior probabilities.

Bayes's theorem applies to the more general case of *k* hypotheses, and then Equation 2.8 becomes

$$\Pr(H_i \mid \text{data}) = \frac{\Pr(\text{data} \mid H_i)\Pr(H_i)}{\sum_{j=1}^{k}\Pr(\text{data} \mid H_j)\Pr(H_j)}. \tag{2.10}$$

When there are more than two hypotheses, it is not obvious how LRs should be scaled; what should be in the denominator? A reasonable choice could be to divide by the hypotheses corresponding to the unrelated, alternative, denoted H_1. Let $\text{LR}_{j,1}$ denote the LR when hypothesis j is compared with 1. *Note*: $\text{LR}_{1,1} = 1$. The general version of Equation 2.9 corresponds to the following formulation of Bayes's theorem:

$$\Pr(H_i \mid \text{data}) = \frac{\text{LR}_{i,1}\Pr(H_i)}{\sum_{j=1}^{k} \text{LR}_{j,1}\Pr(H_j)},$$

which simplifies to

$$\Pr(H_i \mid \text{data}) = \frac{\text{LR}_{i,1}}{\sum_{j=1}^{k} \text{LR}_{j,1}}$$

for flat priors.

2.3.5 BAYES'S THEOREM IN ODDS FORM

Bayes's theorem may alternatively be formulated in odds form:

$$\frac{\Pr(H_1 \mid \text{data})}{\Pr(H_2 \mid \text{data})} = \frac{\Pr(\text{data} \mid H_1)}{\Pr(\text{data} \mid H_2)} \times \frac{\Pr(H_1)}{\Pr(H_2)}, \tag{2.11}$$

which may be formulated verbally as follows:

$$\text{posterior odds} = \text{LR} \times \text{prior odds}.$$

In the above example with flat priors, the prior odds is 1 and the posterior odds equals LR, but the interpretation differs. The posterior odds refers to the probability of the hypotheses given the data: the probability that the alleged father is the father is 4.42 times greater than the probability that an unrelated man is the father. Recall that the LR pertains to the data given the hypotheses.

Example 2.4 Prior odds in the Romanov case. This example is based on the Romanov case and is included here as Bayes's theorem is used with an informative prior. DNA analysis played an important role as documented in [35] and subsequent papers to identify Tsar Nicholas II, Tsarina Alexandra, and three of their five children. The identification used autosomal STR markers to determine the relationship between the mentioned Romanovs found in a grave in Ekaterinburg 1991 and mtDNA to demonstrate that the royal family had been found by comparison of mtDNA with that from known relatives such Prince Philip, Duke of Edinburgh. Two hypotheses were addressed in [35]: "the group is the Romanov family" (H_1) and "the group is an unknown family unrelated to the Romanovs" (H_2). There was evidence a priori that the family was aristocratic. The dental fillings were made of gold and platinum. Furthermore, the age and sex of the bodies appeared correct, and the location of the grave appeared right. On the basis of this evidence, a modest prior odds of 10 was assumed. The LR of 70 then translates to a posterior odds of $70 \times 10 = 700$ according to Equation 2.11. A more detailed statistical analysis is given in [4].

2.4 MUTATION

2.4.1 BIOLOGICAL BACKGROUND

A mutational event brings some change to the genome of an individual. It may occur on the somatic level, impacting only on the individual level, or in the sex cells, affecting future generations. We are mostly interested in the latter. There are several different causes for mutations, including radiation, dysfunctional DNA repair enzymes, and environmental factors. For STR markers, another mechanism for mutations is observed. The effect is called DNA strand slippage error [36], and occurs during DNA replication when the polymerase that duplicates the DNA slips, possibly because of the repeated structures of the STR markers, to produce a new variant with one repetition more or less than the original allele [37]. The probability of observing a variant further away from the original allele, in terms of repeats, decreases fast. The process is illustrated in Figure 2.4.

The slippage error is in fact quite common, compared with "normal" mutations, occurring in roughly 0.5% of all DNA replications.

The mutation rates for forensic loci are relatively high, otherwise these genetic markers might not have been so polymorphic—that is, contain so many alleles. Assume a laboratory handles 1000 paternity cases every year. For each case, something like 16 loci are considered. If the mutation rate is set to 0.001, which is not totally unrealistic,[2] we would expect $1000 \times 16 \times 2 \times 0.001 = 32$ mutations annually, demonstrating that mutations cannot be ignored.

2.4.2 MUTATION EXAMPLE

So far only two markers have been considered. For the third marker, D6S474, the genotypes are 14/15 for the alleged father and 16/17 for the child, and so there is no allele sharing. A direct calculation along the lines considered so far would give an

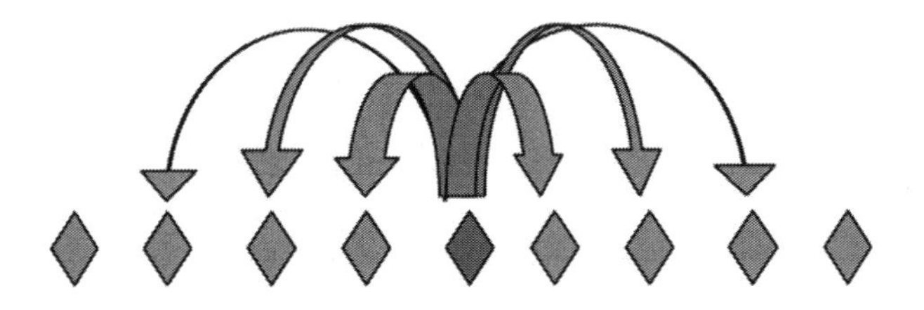

FIGURE 2.4

Probability of mutation decreases with distance.

[2] http://www.cstl.nist.gov/strbase/mutation.htm.

Table 2.2 Mutation Matrix for the Equal Model with Mutation Rate $R = 0.001$

	13	14	15	16	17	18
13	0.9990	0.0002	0.0002	0.0002	0.0002	0.0002
14	0.0002	0.9990	0.0002	0.0002	0.0002	0.0002
15	0.0002	0.0002	0.9990	0.0002	0.0002	0.0002
16	0.0002	0.0002	0.0002	0.9990	0.0002	0.0002
17	0.0002	0.0002	0.0002	0.0002	0.9990	0.0002
18	0.0002	0.0002	0.0002	0.0002	0.0002	0.9990

LR of 0 for this marker. Since the combined LR is obtained by multiplying the results for the individual markers, the overall LR would also be 0. A reasonable approach is to include the possibility of mutation. This can be done in several ways; there are different models. The simplest is to specify a mutation rate—say, $R = 0.001$, or 0.1%—and assume that all mutations are equally likely.

A mutation model is completely specified by the mutation matrix. For this model, the mutation matrix is given in Table 2.2. The diagonal elements $0.999 = 1 - 0.001$ are the probabilities that no mutation happens, while the off-diagonal elements give the mutation probabilities. For instance, allele 16 is passed on as allele 16 with probability 0.999 and as another allele with probability 0.0002. More formally, we describe the mutation model in terms of a transition matrix $M = [m_{ij}]$, where m_{ij} denotes the probability that allele i is inherited as allele j $(i,j = 1,\ldots,n)$. In principle, all numbers of the transition matrix can be specified. In practice this is not a feasible option as the number of parameters would be too large. For instance, for a marker such as SE33 with around 50 different alleles, $50 \times 50 = 2500$ numbers would be needed (or actually a slightly smaller number of freely varying parameters as discussed below). Therefore, parametric models are introduced. Such models should

- be mathematically consistent,
- be formulated in terms of parameters that can be interpreted,
- be biologically reasonable.

The general discussion of models is given in Chapter 6. Here we consider only an example. For instance, in Table 2.2, $m_{17,17} = 0.999$, while

$$m_{17,13} = m_{17,14} = m_{17,15} = m_{17,16} = m_{17,18} = 0.0002.$$

Obviously $0 \leq m_{ij} \leq 1$ and $\sum_{j=1}^{n} m_{ij} = 1$. The latter sum expresses the probability that allele i must end up as some allele.

Table 2.3 Mutation Matrix for the "Stepwise (Unstationary)" Model with Mutation Rate $R = 0.001$ and Range $r = 0.5$

	13	14	15	16	17	18
13	0.99900	0.00052	0.00026	0.00013	0.00006	0.00003
14	0.00035	0.99900	0.00035	0.00017	0.00009	0.00004
15	0.00015	0.00031	0.99900	0.00031	0.00015	0.00008
16	0.00008	0.00015	0.00031	0.99900	0.00031	0.00015
17	0.00004	0.00009	0.00017	0.00035	0.99900	0.00035
18	0.00003	0.00006	0.00013	0.00026	0.00052	0.99900

Recalling that the alleged father is 14/15 and the child is 16/17 for the marker D6S474, we find there are four possible mutations:

$$14 \to 16,\ 14 \to 17,\ 15 \to 16,\ 15 \to 17.$$

The equal model unreasonably specifies these mutational events as equally likely. The shortest mutation, $15 \to 16$, should be the most probable, and the longest mutation, $14 \to 17$, should be the least likely as illustrated in Figure 2.4. There is, therefore, a need for more biologically plausible models, and several are available. We consider only the simplest version of the "Stepwise" model discussed further in Chapter 6 and [38, 39]. For this model, the relative probabilities of different mutations are specified. Specifically, the mutation range r is the probability of a one-step mutation divided by the two-step mutation probability. The mutation matrix for a stepwise model with $R = 0.001$ and $r = 0.5$ is given in Table 2.3. So with $r = 0.5$, the mutation $13 \to 15$ is half as likely as the mutation $13 \to 14$, or numerically $0.00026/0.00052 = 0.5$. For a three-step mutation, $13 \to 16$, compared with a one-step mutation, we find $0.00013/0.00052 = 0.25 = r^2$.

2.4.3 MUTATION FOR DUOS

Typically, the numerical calculations require a computer program even in quite simple cases involving mutations. An exception is the parent-child case for which there is the general formula

$$\mathrm{LR} = \frac{1}{4}\frac{(m_{ac} + m_{bc})p_d + (m_{ad} + m_{bd})p_c}{p_c p_d}. \tag{2.12}$$

The parent genotype is a/b and the child genotype is c/d, and as mentioned previously, m_{ij} denotes the probability that allele i ends up as j.[3] For the "equal"

[3]There are no restrictions on alleles a, b, c and d; they may or may not differ. The formula is presented in [110] without proof; see Equation 6.14 for a way to derive this formula.

model, $m_{i,j} = m$ if i differs from j. Therefore, if the alleged father and the child share no alleles,

$$\mathrm{LR} = \frac{1}{2}\frac{m(p_c + p_d)}{p_c p_d}.$$

For the marker D6S474, $a = 14$, $b = 15$, $c = 16$, $d = 17$, $p_c = 0.25$, $p_d = 0.097826$ and $m = 0.002$ which gives the likelihood ratio

$$\frac{1}{2}\frac{0.002(0.250000 + 0.097826)}{0.250000 \times 0.097826} = 0.0014.$$

The calculations below are based on the stepwise model specified by the mutation matrix in Table 2.3 and Equation 2.12, and give

$$\frac{1}{4}\frac{(0.00017 + 0.00031) \times 0.250000 + (0.00009 + 0.00015) \times 0.097826}{0.250000 \times 0.097826} = 0.001.$$

2.4.4 DEALING WITH MUTATIONS IN PRACTICE

When using a computer program for computations, a marker may be defined by the minimum set of alleles needed or by including all alleles. For instance, if only alleles A and B are observed in the case data, only these alleles are needed in addition to a rest allele with frequency defined to ensure that all allele frequencies add to 1. We refer to the resulting model as "minimal" since the minimum number of alleles is specified. On the other hand, all alleles of the marker can be entered. For all models involving mutations, slightly different results will be obtained depending on whether a minimal specification is used or not. This is a general feature of any mutation framework as the different setups give rise to different models. In our example, the LR for D6S474 was calculated on the basis of a model including all alleles; the result would change for a minimal specification. Generally, we recommend use of the complete database.

We recommend specifying mutation models generally before considering the data of a specific case and therefore routinely specify mutation models for all markers. The alternative may appear strange as then the model is changed once an inconsistency has been observed. This could be considered dubious as the model is changed to fit a specific observation. The disadvantage of routinely accommodating mutations is that calculations become more complicated. It is no longer easy to check values except for simple cases such as the duo case we have discussed.

It may be prudent to compute with several mutation models to test the robustness of the conclusions, and even to be able to operate with different mutation rates for paternal and maternal alleles. The models implemented in `Familias 3` are described in greater detail in the manual.

2.5 THETA CORRECTION

Population stratification and *relatedness* may invalidate previous calculations. For instance, HWE may not apply in paternity cases. The parents may be remotely related, although not knowingly, simply because they belong to the same subpopulation. Balding and Nichols [40] proposed a practical way of handling these problems that has been endorsed in the recommendations of the National Research Council [41]. The approach may be given a genetic or evolutionary argument as well as a more statistical one along the lines described in Section 6.1.2. We give only a heuristic argument in the next section leading to a computational formula.

2.5.1 SAMPLING FORMULA

The effect of population stratification is modeled by the *coancestry coefficient* $\theta \in [0, 1]$. The value 0 corresponds to HWE, whereas positive values increase the probability of homozygotes, as will be illustrated below. Assume alleles are sampled sequentially. The probability of sampling allele i as the first allele is p_i. Suppose i is sampled as the j'th allele and let b_j denote the number of alleles of type i among the $j-1$ previously sampled alleles. To achieve that the j'th allele is sampled as a weighted compromise between sampling from the set of already sampled alleles, and sampling with the allele frequencies p_i, we use $[\theta b_j + (1-\theta)p_i]/[\theta(j-1) + (1-\theta)]$ as the probability for sampling allele i. Rearranging, this gives the sampling formula

$$p_{i'} = \frac{b_j\theta + (1-\theta)p_i}{1 + (j-2)\theta}. \tag{2.13}$$

Note: If $\theta = 0$, this probability remains p_i, as is reasonable. Furthermore, if $b_j/(j-1) > p_i$, more alleles of type i are sampled than expected, and the probability that the next will be of type i exceeds p_i if $\theta > 0$. If we sample two alleles, it follows from Equation 2.13 that both alleles are of type i with probability $\theta p_i + (1-\theta)p_i^2$. This expression reduces to the probability for the HWE probability for a homozygote when $\theta = 0$. Similarly, we find that the probability of a heterozygote of type A/B equals $2(1-\theta)p_A p_B$, which again coincides with the HWE probability. Observe that coancestry leads to an exceedance of homozygotes and fewer heterozygotes as can be seen from modified versions of Equations 2.1 and 2.2 with $\theta = 0.01$:

$$\Pr(G_{\mathrm{AF}} = 17/18) = 2p_{17}p_{18}(1-\theta) = 0.0563\ (0.0569),$$

$$\Pr(G_{\mathrm{AF}} = 8/8) = \theta p_8 + (1-\theta)p_8^2 = 0.3093\ (0.3068),$$

where the numbers in parentheses are for $\theta = 0$.

Example 2.5 Theta correction. For our continued example, $p_{17} = 0.204$ and we found $\mathrm{LR}_1 = 1/(2p_{17}) = 2.45$ for D3S1358. Using Equation 2.13, we have

$$\mathrm{LR} = \frac{(1+2\theta)}{2\times(2\theta+(1-\theta)p_{17})} = 2.30$$

with $\theta = 0.01$. Values of θ between 0.01 and 0.03 have been recommended [32]. Larger values can be appropriate in more extreme cases.

2.6 SILENT ALLELE

There are cases when alleles fail to amplify. If this relates to sequence variations in the flanking regions of the STR marker, the problem is known as a *silent allele*. Using alternative methods, one may detect a silent allele, and for this reason Gill and Buckleton [14] recommends the term "null alleles" not be used. The silent allele frequency and frequency of the other allele frequencies should add to 1. Information on frequencies of silent alleles is given at http://www.cstl.nist.gov/strbase/NullAlleles.htm. An example involving silent alleles is provided in Exercise 2.11. Mutation, theta correction, and other factors are ignored in this section to illustrate the effect of silent alleles. Let S denote the silent allele. There are three alleles involved with frequencies, say, $p_A = 0.2$, $p_B = 0.2$, and $p_S = 0.05$. The probability of the genotype A/− is calculated as $p_A^2 + 2p_Ap_S = 0.06$.

Example 2.6 Silent allele. The alleged father and the child are both 8/8 for the marker TPOX in our returning example. Suppose there is a silent allele with frequency $p_S = 0.05$. An adjustment of allele frequencies is necessary since these should sum to 1 with the silent allele included. We keep the allele frequency $p_8 = 0.5539$. Then

$$\begin{aligned}
\Pr(\text{data} \mid H_1) &= \Pr(G_{\mathrm{AF}} = 8/8, G_{\mathrm{CH}} = 8/8 \mid H_1) \\
&\quad + \Pr(G_{\mathrm{AF}} = 8/S, G_{\mathrm{CH}} = 8/8 \mid H_1) \\
&\quad + \Pr(G_{\mathrm{AF}} = 8/8, G_{\mathrm{CH}} = 8/S \mid H_1) \\
&\quad + \Pr(G_{\mathrm{AF}} = 8/S, G_{\mathrm{CH}} = 8/S \mid H_1) \\
&= p_8^3 + p_8^2p_S + p_8^2p_S + p_8^2p_S + p_8p_S^2
\end{aligned}$$

and

$$\mathrm{LR} = \frac{p_8(p_8^2 + 3p_8p_S + p_S^2)}{(p_8^2 + 2p_8p_S)^2} = 1.66.$$

2.7 DROPOUT

In recent years the field of forensic genetics has been introduced to increasingly sensitive methods that allow samples with very small amounts of DNA to be analyzed. This necessitates the handling of complicating factors such as allelic dropout. In statistical parlance, dropout is similar to missing data: there are some alleles that fail to amplify.

For mixtures in criminal cases, degraded DNA is more commonly encountered, and there is a tradition for handling dropout in a probabilistic framework. The International Society of Forensic Genetic DNA Commission recommends the use of a probabilistic approach which includes dropout for forensic case work [42] and also give guidance on how to estimate the dropout probability.

For cases based on reference samples of good quality, dropout is not an issue. We rather aim to describe methods appropriate for investigations involving missing persons and disaster victim identification, and maybe particularly archeogenetic analyses. Partial profiles can lead to falsely regarding markers as homozygous, whereas excluding problematic markers could cause loss of valuable information and biased results. Silent alleles and mutation can sometimes be alternative explanations for apparently inconsistent findings, and they can be modeled jointly as exemplified in Exercises 2.15 and 2.16. Mutations and silent alleles are inherited, as opposed to dropouts.

Less attention has been given to dropouts in kinship cases, the topic addressed below. Some previous work [43] (and the associated software DNA-view; http://dna-view.com/dnaview.htm) assigns an LR of 0.5 to loci with a possible dropout as a rough estimate. An implementation of dropout for kinship cases can be found in the Bonaparte software tool [44, 45].

The model implemented in `Familias` is described in [46]. This model extends the previously mentioned work by building on [47, 48] and is summarized in Example 6.10. Here, we only mention that each allele has a fixed probability d for not being observed, and whether or not an allele is observed is independent between alleles.

2.8 EXCLUSION PROBABILITIES

Section 8.1.1 expands on the introduction below. In some situations—for example, in paternity cases—part of the population can be excluded from being the *person of interest*. As the father and the child should share an allele on every locus (disregarding mutations and other artifacts such as silent alleles and dropout), using the allele frequencies, one can calculate what fraction of the general population cannot be excluded. This amounts to finding the probability of exclusion (PE), or equivalently the random man not excluded (RMNE) probability: RMNE $= 1 -$ PE. Observe that PE can be determined before any persons have been genotyped in a case. Such prior calculations characterize the power in a meaningful way. Ideally, the PE should be

Table 2.4 The Joint Distribution of G_{AF} (Rows) and G_{CH} if there is no Relationship

	(a/a)	(b/b)	(a/b)
(a/a)	0.0625	0.0625	0.1250
(b/b)	0.0625	0.0625	0.1250
(a/b)	0.1250	0.1250	0.2500

close to 1, and if this is not the case, an effort can be made to obtain more data. Additional persons may be approached for genotyping, or extended marker sets can be analyzed.

The calculations are most easily demonstrated for a SNP marker. Table 2.4 shows the genotype probabilities for the alleged father and child when they are unrelated and the allele frequencies are 0.5 and 0.5. From the table we realize that exclusion occurs if the individuals are homozygous for different alleles, and therefore $\text{PE} = 2 \times 0.0625 = 0.1250$. For a marker with three alleles, exclusion would also happen if one individual is heterozygous and the other individual is homozygous for a different allele. With four or more alleles, both individuals could be heterozygous for different alleles. On the basis of the above reasoning, we realize that for a parent-child relationship

$$\text{PE} = \sum_{i=1}^{n} p_i^2(1-p_i)^2 + 2\sum_{i<j} p_i p_j (1-p_i-p_j)^2. \tag{2.14}$$

The first term corresponds to the alleged father being homozygous, while the latter term corresponds to the alleged father being heterozygous but with no alleles shared with the child. In [49] a general approach to calculating RMNE (meaning that there are no restrictions on the pedigrees involved) is presented. There are formula available for other specific cases, as summarized in [15].

In our continued example, $\text{PE}_1 = 0.4156$ for the marker D3S1358. (The calculations are explained in Exercise 5.6.) In other words, the probability that a random man is excluded as a father before any data has been obtained is 0.4156. Equivalently, there is a probability of $\text{RMNE}_1 = 1 - 0.4156 = 0.5844$ that he is not excluded. For TPOX, $\text{PE}_2 = 0.2112$ and $\text{RMNE}_1 = 1 - 0.2112 = 0.7888$. The probability of at least one exclusion is

$$1 - 0.5844 \times 0.7888 = 0.5390. \tag{2.15}$$

It is essential that the calculation is based on the complete database specifying all alleles and their frequencies. The PE will be close to 1 if the number of alleles n is large and the allele frequencies are $1/n$. Similarly, the probability of not excluding a random man will be close to 0 for a realistic number of markers—say, 16.

2.8.1 RANDOM MATCH PROBABILITY

The framework described thus far has dealt with relationship inference between individuals. The concepts may be extended to also include direct sample comparisons. In disaster victim identification, described in detail in Chapter 3, it is not uncommon to obtain reference samples from, for instance, the razor blade or toothbrush of a person to be used for subsequent identification. It is then convenient to also have a measure of the evidence given that two profiles are found to be identical.

In forensic genetics the random match probability (RMP) is frequently used to state how probable a certain profile at a crime scene is. The inverse of this measure, 1/RMP, is a version of the LR. We may formally define the following:

H_1: Two profiles (G1 and G2) originate from the same individual.
H_2: The two profiles originate from two unrelated individuals.

An elaboration of these ideas is presented in Section 3.3.2. To illustrate, consider two genetic profiles, both homozygous 17/17 for the marker D3S1358. Then 1/RMP is simply computed as $1/0.204^2 \approx 24$, indicating the two profiles are 24 times likelier to originate in the same individual. The RMP, $0.204^2 \approx 0.04$, is the probability that a random man will have this profile in the population.

2.9 BEYOND STANDARD MARKERS AND DATA

So far we have discussed methods based on standard forensic markers: unlinked autosomal STRs in linkage disequilibrium. Other markers are available and can be appropriate for specific applications. Here we briefly comment on marker data which require modification of the methods presented so far.

Obviously, changing from STRs to SNPs involves only a simplification, and therefore does not warrant further discussion.

2.9.1 X-CHROMOSOMAL MARKERS

The relevance of the X chromosome for relationship inference has been accentuated in recent years, and a large number of papers have appeared ([50] is an early publication). The particular inheritance pattern with recombination only for females makes such markers particularly useful for some specific cases. A standard example is to distinguish between paternal and maternal female half siblings on the basis of genotype data from the two girls. For this case, autosomal markers provide no information whereas X-chromosomal markers are powerful. There are, however, few completely independent markers on the X chromosome. For this reason, we refer the reader to Chapter 4, which presents models and the `FamLinkX` implementation taking linkage and linkage disequilibrium into account. Likelihood calculations that use the `R` package `paramlink` are exemplified in Exercise 8.5.

2.9.2 Y-CHROMOSOMAL AND mtDNA MARKERS

Lineage markers like those on the Y chromosome and mtDNA, have been proved important for a number of applications. In forensics, such markers have been very useful to deal with distant relationship inference spanning several generations, as was exemplified in the Jefferson case [51]. The Y chromosome is passed on from father to sons, and can therefore be treated as one marker. In this sense, it is easy to deal with Y-chromosomal markers. Mutations can complicate matters. However, most of the discussion in the scientific literature has focused on the estimation of haplotype frequencies and how to deal with match probabilities when the haplotype in a case has not previously been observed. Caliebe et al. [52] emphasize that there is no simple approach. We return to the problem in Chapter 4 with a Bayesian approach to inference for haplotype frequencies. Exercise 5.7 exemplifies several approaches, including methods proposed in [53, 54].

Databases are needed to estimate Y-haplotype frequencies. YHRD (http:/yhrd.org) is a Y-chromosomal STR searchable database that allows users to estimate haplotype frequencies.

The Romanov case [35] exemplifies the utility of mtDNA markers. The mtDNA is passed on from a mother to all children. As for Y-chromosomal markers, mtDNA is treated as a single marker for forensic calculations. Guidelines for mtDNA typing are provided in [55]. The Innsbruck Institute of Legal Medicine has developed the forensic mtDNA population database EMPOP (http://empop.org). This site provides information on palaeogenetic, medical genetic, and forensic genetic investigations. In particular, the abundance of a specific mtDNA sequence can be estimated from the appropriate database.

2.9.3 DNA MIXTURES

DNA mixture evidence refers to data where several individuals may have contributed to a biological stain. Rape cases present an important example, and occur frequently in forensic case work. Typically, the DNA profile based on a vaginal swab will indicate the presence of the victim and one or more men; the *electropherogram* typically displays more than two peaks. As this book deals with relationship inference, for which DNA profiles are normally based on reference profiles from single individuals, mixture evidence will play no large part. There are, however, cases where data comes in the form of mixtures, and these are discussed further in Section 3.4.2.

2.10 SIMULATION

For a specified model and hypothesis, simulated values $\text{LR}_1, \ldots, \text{LR}_{\text{Nsim}}$ can be generated. From these values, summary statistics such as the mean and standard deviation can be estimated, as can the exceedance probability $\Pr(\text{LR} > t \mid H)$

for a prescribed threshold t. A relevant value for t corresponds to thresholds given in published tables (there are several) relating LR values to verbal statements. For instance, it has been suggested that values of the LR exceeding 100,000 translate to "very strong." Exercise 2.17 exemplifies simulation in `Familias`. A more complete discussion of simulation is given in Chapter 8.

2.11 SEVERAL, POSSIBLY COMPLEX PEDIGREES

In human genetics, *complex pedigrees* involve inbreeding. The terminology differs in forensic genetics. "Complex" indicates something more complicated than the standard trio, but the term is not precisely defined. In our applications, the pedigree could both involve inbreeding and be large. However, pedigrees involving more than, say, 10 individuals are rarely encountered, Exercise 4.10 provides an exemption.

In previous sections, only two alternatives were considered. However, sometimes more alternatives need to be considered. The basic calculation leading to the likelihood for a given hypothesis extends directly. However, a choice needs to be made as far as the LR is concerned as the definition intrinsically involves only two pedigrees. All hypotheses can, for instance, be scaled against a common alternative. The number of LRs is one less than the number of hypotheses. If scaling is done against the least likely alternative, all LRs exceed 1. This will not be the case for other scaling alternatives, and it is important to factor this in the reference hypothesis when interpreting the calculations. Exercise 2.6 illustrates inbreeding and more than two hypotheses as both inbreeding by the father and inbreeding by the brother of the mother are considered in addition to an unrelated man being the father. Below we extend our continued example to three alternatives.

Example 2.7 Posterior probabilities for relationships. Consider once more the marker D3S1358. The hypotheses are that the alleged father is the father of the child (H_1), they are unrelated (H_2), or they are brothers (H_3). The likelihoods for these alternatives are

$$\begin{aligned} L_1 &= L(H_1) = p_{17}^2 p_{18}, \\ L_2 &= L(H_2) = 2p_{17}^3 p_{18}, \\ L_3 &= L(H_3) = 0.5 \times p_{17}^3 p_{18} + \frac{1}{2} p_{17}^2 p_{18}. \end{aligned}$$

The LR in Equation 2.3 is L_1/L_2, whereas the LR in Equation 2.7 is L_3/L_2, i.e., the unrelated alternative is chosen as a reference, but there are other options. The need to single out the denominator disappears if the posterior probabilities are reported, but then a prior is needed. Assuming a flat prior, we have

$$\Pr(H_i \mid \text{data}) = \frac{L_i}{L_1 + L_2 + L_3},$$

which gives posterior probabilities 0.4975, 0.2030 and 0.2995 (with $p_{17} = 0.2040$ and $p_{18} = 0.1394$) for the three alternatives. *Note*: The LRs can be retrieved from these posteriors as a flat prior has been invoked. For instance $L_1/L_2 = 0.4975/0.2030 = 2.45$ as calculated previously in Equation 2.3.

2.12 CASE STUDIES

2.12.1 PATERNITY CASE WITH MUTATION

Previously in our returning example summarized in Table 2.5, we used some of the markers (D3S1358, D6S474, and TPOX), but now all markers will be taken into

Table 2.5 Genotype Data for a Child and Alleged Father (AF) Along with LRs

System	Child	AF	LR	LR (mut)
D3S1358	17/17	17/18	2.450	2.449
TPOX	8/8	8/8	1.805	1.804
D6S474	16/17	14/15	0.000	0.001
TH01	6/9	6/7	1.195	1.194
D21S11	29/30	28/29	1.096	1.095
D18S51	14/16	16/17	2.153	2.152
PENTA_E	7/11	11/16	2.408	2.406
D5S818	12/12	12/13	1.406	1.405
D13S317	8/8	8/11	4.042	4.038
D7S820	9/10	9/13	1.434	1.432
D16S539	13/14	11/14	8.312	8.305
CSF1PO	10/10	10/11	2.025	2.023
PENTA_D	8/11	8/13	11.989	11.978
VWA	19/19	17/19	5.565	5.560
D8S1179	13/16	11/16	9.651	9.641
FGA	21/22	21/21	2.956	2.954
D12S391	19/22	19/23	2.184	2.182
D1S1656	14/16	14/15	3.333	3.331
D2S1338	18/20	18/23	3.147	3.144
D22S1045	12/12	12/15	26.748	26.72
D2S441	10/13	10/15	1.446	1.446
D19S433	12/15	12/14	3.344	3.340
Total			0	25,070,642

Notes: *The rightmost column is based on a stepwise unstationary mutation model with mutation rate 0.001 and range 0.5 for all markers.*
All information required to reproduce this table is provided as http://familias.name/Table2.5.fam

account. Initially, the calculations are for the same standard hypotheses, paternity (H_1) versus nonpaternity (H_2), and it is for these alternatives that LRs are given in the Table 2.5. Obviously, the overall LR remains 0 if mutations are not considered when the number of markers is extended from 3 to 22.

The incompatibility between the alleged father and the child for D6S474 is the interesting issue to discuss. If possible, further markers could be genotyped, but such extended data is not available. Dropout could have been an option to consider if one of the typed individuals had been homozygous, but this is not the case, and dropout is therefore not further discussed. Dropout is also a priori not likely if the reference values are of good quality. For the same reason, we do not consider genotyping error.

In this case, with 21 out 22 markers being compatible with paternity, it is reasonable to explain the finding by a mutation; see Section 2.4. As mentioned previously, we recommend the use of a mutation model routinely for all markers. However, as we can see, the mutation model has little impact except for D6S474. The mutation parameters typically differ between markers and also differ between females and males, but for simplicity we have used $R = 0.001$ and $r = 0.5$ as before. One could try different mutation models and different parameter values, but unless extreme choices are made, the LR will remain extremely large and the conclusion will remain unchanged.

Introducing an alternative hypothesis involving a close relative of the alleged father as the true father could however be a plausible alternative explanation. Normally, such an alternative explanation would have to be requested by the party that requested the analysis in the first place. Nonetheless, we have calculated for the brother alternative (H_3) and found

$$\Pr(H_1 \mid \text{data}) = 0.9913, \quad \Pr(H_2 \mid \text{data}) = 3.954 \times 10^{-8}, \quad \Pr(H_3 \mid \text{data}) = 0.0087,$$

where a flat prior of $1/3$ for each of the three hypotheses was used. The LR comparing paternity with the brother is therefore $0.9913/0.0087 = 114.0$.

2.12.2 WINE GRAPES

With the exception of the case study discussed next, the examples in this book are dedicated to human applications. There are a number of nonhuman applications, including "wildlife forensic science," as pointed out in [56]. The case discussed below, however, does not relate to either wildlife or forensics, but relates rather to plant genetics or more specifically relationship inference for wine grapes.

We will use some of the wine data from [1], which was also discussed in Section 4.3 in [10]. The summary of [1] starts as follows: "The origins of the classic European wine grapes (*Vitis vinifera*) have been the subject of much speculation. In search of parental relationships, microsatellite loci were analyzed in more than 300 grape cultivars." To underline the importance of this research, Bowers et al. [1, page 1564] state the following: "Knowledge of parental relationships as those reported here can facilitate rational decisions regarding the size of grape germplasm core collections, which are constantly threatened by economic constraints."

Table 2.6 Excerpts from Table 1 in [1]

	Genotype			
Locus	**P**	**C**	**G**	**Observed**
VVMD28	221	221	231	0.057637
	239	231	249	0.115274
VVS2	137	137	133	0.040346
	151	143	143	0.171470
VVMD31	216	214	212	0.086455
	216	216	214	0.214697
VrZAG79	239	243	237	0.094697
	245	245	243	0.108696

Notes: *The genotypes of Chardonnay (C) and its assumed parents Pinot (P) and Gouais blanc (G) are shown for four loci. The "Observed" column gives the estimated allele frequencies of the corresponding progeny allele, the one in column "C," based on 322 genotypes. For instance, the frequency of allele 221 for locus VVMD28 is 0.057637 (a large number of digits have been included to allow for comparison of calculations with [1]).*

Excerpts of the data in [1] are shown in Table 2.6. The genotypes of Chardonnay (C) and its assumed parents Pinot (P) and Gouais blanc (G) are shown for four loci. We assume there is HWE and the LRs can be multiplied. Moreover, complicating factors such as mutations and dropout will not be considered. There is one difference between this example and human examples: the sex of wine grapes is not an issue ("most grape cultivars are hermaphrodic" [1, page 1564]). The software we know requires sex to be specified. This does not affect the likelihoods, but it does have implications for how pedigrees are generated and handled in the software. Assume P is specified as a male, and G and C are specified as females. Then the software allows calculations for pedigrees where P and C have offspring. In this wine grape application, one could imagine offspring from G and C as well. Technically this can be handled by the software by introducing a copy of G (or C) of the opposite sex.

Prior model for wine grapes

Unfortunately, we do not have the expertise to specify sensible priors, but we may still use this data for exemplification. According to Bowers et al. [1, page 1564], "grape is intolerant of inbreeding" and so we include no inbred pedigrees. Figure 2.5 shows nine possible pedigrees, including the one where P and G are parents of C, which was found to be the likeliest in [1].

These pedigrees are also listed in Table 2.7, and a flat prior of 1/8 is prescribed for each of them.

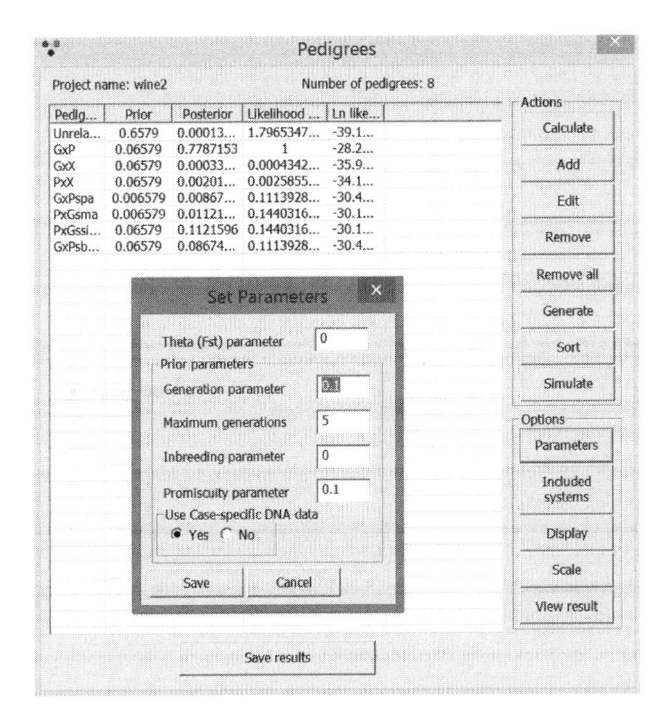

FIGURE 2.5

Wine example with prior. The parameters under "General settings" only affect priors and posteriors and not LR and are explained in the manual for `Familias`.

Table 2.7 Results of the Wine Example

	Pedigree	Prior	Likelihood	Posterior
1	Unrelated	0.125	1.00994×10^{-17}	0.000012
2	G × P	0.125	5.62162×10^{-13}	0.660551
3	P × X	0.125	1.45349×10^{-15}	0.001708
4	G × X	0.125	2.44133×10^{-16}	0.000287
5	P × Gsma	0.125	8.09692×10^{-14}	0.095140
6	P × Gssister	0.125	8.09692×10^{-14}	0.095140
7	G × Pspa	0.125	6.26209×10^{-14}	0.073581
8	G × Psbrother	0.125	6.26209×10^{-14}	0.073581

Notes: *The leftmost column shows the pedigree and, for instance, G × P indicates that P and G are parents of C. For P × X, only the identity of P is assumed to be known. Alternatives 5 and 6 (and also 7 and 8) are not distinguished in [1]. However, in the presence of prior information, the evidence for these pedigrees may differ. To explain the notation, note that the last pedigree corresponds to G and a brother of P being parents.*

Some pedigrees—for instance the two last in Table 2.7—cannot be distinguished—that is, likelihoods coincide. Prior information modeled by means of our approach may serve to distinguish such alternatives, as discussed in [10].

Likelihoods for wine grapes

The LR obtained by dividing the likelihood (see Table 2.7) of the data if the parents of C are G and P by the likelihood if the parents are G and some unknown, denoted X, becomes

$$\frac{\Pr(\text{data} \mid \text{G} \times \text{P})}{\Pr(\text{data} \mid \text{G} \times \text{X})} = \frac{5.62 \times 10^{-13}}{2.44 \times 10^{-16}} = 2303. \tag{2.16}$$

This number coincides with the results in [1]. By extending the analysis to 32 loci, we find $\text{LR} = 4.36 \times 10^{17}$. The latter calculation requires access to unpublished data. Figure 2.5 is based on 4 markers and the `Familias` file is available as http://familias.name/wine2.fam.

2.13 EXERCISES

Software, input files and solutions (some in the form of videos) can be found following links from http://familias.name.

Exercise 2.1 Simple paternity case. The purpose of this exercise is to illustrate the basic methods and `Familias` with a simple paternity case. Figure 2.6 shows a mother (undisputed), an alleged father and a child.

We consider the hypotheses of a standard duo case as explained in Section 2.3.1, that is

H_1: The alleged father is the biological father.
H_2: A random man is the biological father.

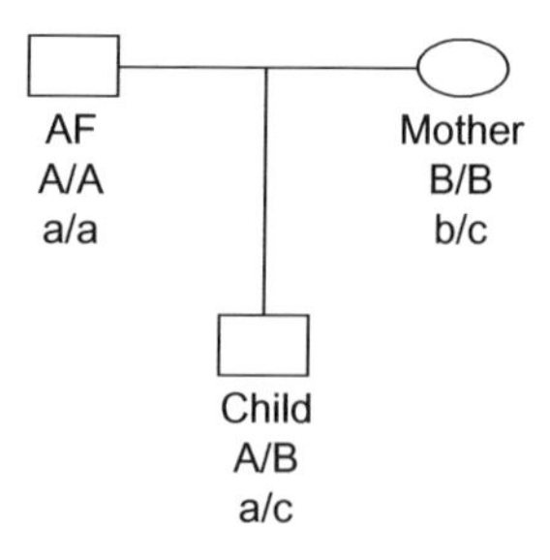

FIGURE 2.6

A standard paternity case with genotypes for two markers; see Exercise 2.1. AF, alleged father.

(a) Consider first only one autosomal locus, called S1, with alleles A, B, and C, as displayed in Figure 2.6. The allele frequencies are $p_A = p_B = 0.05$ and $p_C = 0.9$. Explain why LR $= 1/p_A$. How do you interpret the LR?

(b) Calculate the LR with `Familias`.

(c) There is a second autosomal locus, called S2, with alleles *a*, *b*, *c*, and *d* with allele frequencies 0.1, 0.1, 0.1, and 0.7, respectively and marker data as shown in Figure 2.6. Calculate the LR for this marker and also for the two first markers combined with `Familias`.

(d) Verify the `Familias` answer calculated above using pen and paper.

(e) Generate a report by clicking `Save results` with options `Only report, Rtf file, Complete`. The report includes all input and all output. Check that the report file is correct and contains sufficient information to reproduce the calculations. In particular, check the LR for markers S1 and S2 and the combined likelihood ratio.

(f) Save the `Familias` file (we suggest the file extension fam). Exit `Familias`.

(g) Start `Familias` and read the previously saved file. Calculate 1/RMP for the alleged father by hand and in `Familias`. *Hint*: Mark AF (alleged father) in the `Case-Related DNA data` window and press `Compare DNA`.

(h) We next consider theta (θ) correction. For simplicity we will use only the first marker, S1. The θ parameter is called the kinship parameter in `Familias` and is set with `Options` in the `Pedigrees` window. Set the θ parameter to 0.02. Calculate the LR. To get calculations for selected markers only, in this case S1, use the `Included systems` button. Check that your answer coincides with the following theoretical result:

$$\text{LR} = \frac{1+3\theta}{2\theta + (1-\theta)p_A}. \tag{2.17}$$

(i) Discuss the assumptions underlying the calculations in this exercise.

Exercise 2.2 Simple paternity case with mutation. We consider a duo case (see Figure 2.7) with one marker, VWA.

The allele frequencies are given in Table 2.8. The alleged father is 14/15, the child is 16/17, and the hypotheses are as in Exercise 2.1.

(a) Explain why LR $= 0$. Confirm this answer with `Familias`.

(b) Use the equal probability mutation model with mutation rate $R = 0.007$ for both males and females and calculate LR.

(c) It can be shown (see Exercise 2.8) that

$$\text{LR} = \frac{m(p_{16}+p_{17})}{2p_{16}p_{17}}, \tag{2.18}$$

Use this formula to confirm the `Familias` calculation. Obtain *m*, the probability of mutating to one specific allele; use `File > Advanced` options. Explain the difference between *R* and *m*.

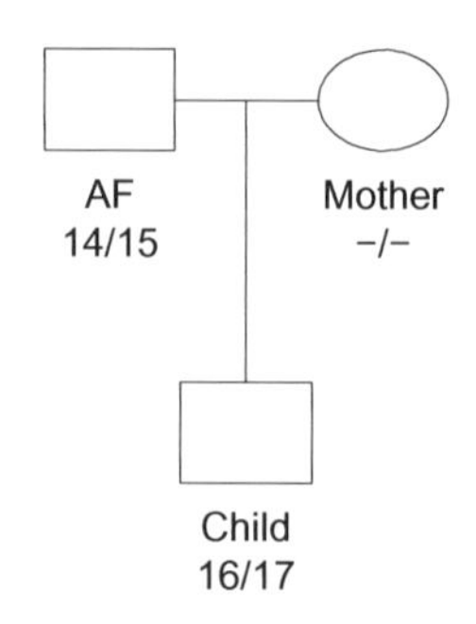

FIGURE 2.7

A paternity case with possible mutation. AF, alleged father.

Table 2.8 Allele Frequencies for Exercise 2.2

Allele	Frequency
14	0.072
15	0.082
16	0.212
17	0.292
18	0.222
19	0.097
20	0.020
21	0.003

Exercise 2.3 Missing sister. The College of American Pathologists has several proficiency testing programs targeted at laboratories that perform DNA typing of STR loci. The following is a test from 2011: Hikers come across human skeletal remains in a forest. Evidence around the site provides a clue as to the identity of the individual. You are asked to test a bone to determine if the individual (bone) is related to an alleged mother and the mother's other daughter, the alleged full sister; see Figure 2.8.

The hypotheses are as follows:

H_1: The bone belonged to the daughter of the alleged mother and the sister of alleged full sister.

H_2: The bone belonged to someone unrelated to the alleged mother and her daughter.

(a) Enter the data manually and calculate the LR. The genotypes and allele frequencies for this marker (F13B) are given in Figure 2.8.

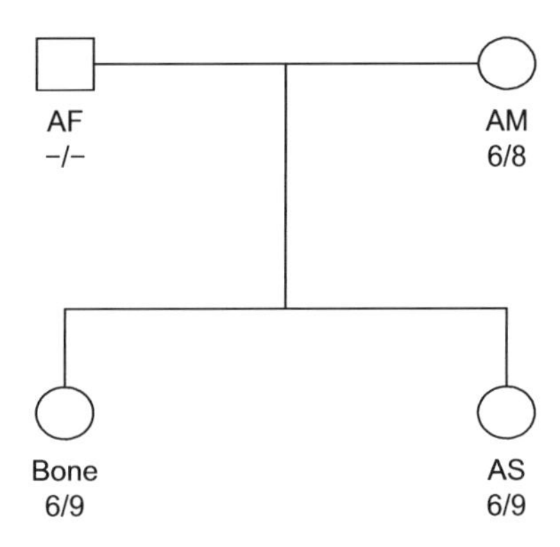

FIGURE 2.8

The case of the missing sister. The alleles (allele frequencies) are 6 (0.086), 8 (0.152), and 9 (0.328) for this marker (F13B). AF, alleged father; AM, alleged mother; AS, alleged full sister.

(b) Read the input from the file `Exercise2_3.fam`. Calculate the LR based on all markers.

(c) Find the LRs of the individual markers. Check that the answer for F13B corresponds to the one you found in problem (a) above.

(d) One of the markers, D7S820, gives a very large LR—namely, 11,189. What is the reason for this large LR? What is the combined LR if marker D7S820 is removed?

Exercise 2.4 Grandfather–grandchild. Two individuals, GF and GS, are submitted to the laboratory for testing. The alternatives are as follows:

H_1: GF is the grandfather of GS.
H_2: The individuals GF and GS are unrelated.

Figure 2.9 shows the pedigree corresponding to H_1 with data for the first marker, D3S1358.

(a) Enter the data given in Figure 2.9 for D3S1358 manually into `Familias` and calculate the LR for the first marker shown in Figure 2.9.

(b) Calculate the LR based on all markers. Read the input from the file `Exercise2_4.fam`.

(c) In the College of American Pathologists exercise it was stated that GF and GS share the same Y haplotype and that the frequency of this haplotype is 0.0025. Can this information be used?

Exercise 2.5 Simple paternity case. (Probability of paternity). We revisit Exercise 2.1. Rather than calculating the LR, we will now calculate the Essen-Möller index W defined as the probability of H_1 conditional on the genotypic data. Assume a priori that hypotheses H_1 and H_2 are equally likely. Then, it can be shown that

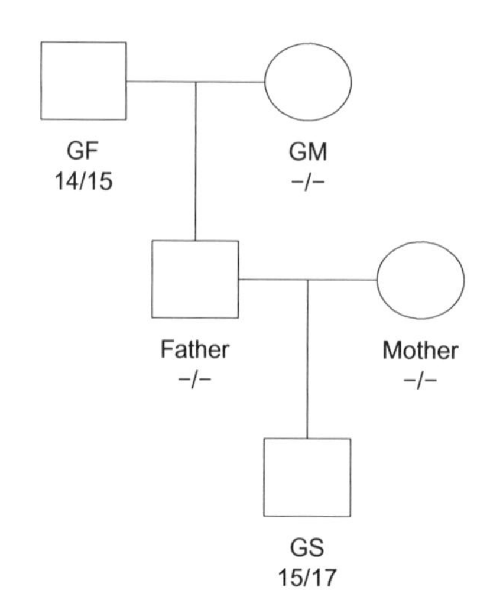

FIGURE 2.9

Grandfather-grandchild. The alleles (frequencies) for D3S1358 are 14 (0.122), 15 (0.258), and 17 (0.197).

$$W = \Pr(H_1|\text{data}) = \frac{\text{LR}}{\text{LR}+1}. \tag{2.19}$$

(a) Recall that LR = 20 for the first marker. Calculate W.
(b) Recall that LR = 200 for two markers. Calculate W.
(c) Use `Familias` with `Exercise2_1.fam` to calculate W for the two cases above.
(d) Do you prefer LR or W?

Exercise 2.6 Inbreeding, LR. (Several alternatives). Consider the following hypotheses:

H_1: The alleged father, the undisputed father of the mother, is also the father of her child.
H_2: A random man is the father of the child.

Figure 2.10 shows the pedigree corresponding to H_1. The allele frequencies are $p_1 = p_2 = p_3 = 0.05$.

(a) Use `Familias` to calculate the LR based on the genotypes given in Figure 2.10. (You are encouraged to enter the data manually, but there is an input file `Exercise2_6.fam` available.)
The defense claims that one should rather consider the hypothesis

H_3: The brother of the mother is the father of the child. See Figure 2.11.

The LR can be calculated in several ways depending on the choice of the reference. Calculate LR (H_1/H_2) and LR(H_1/H_3).

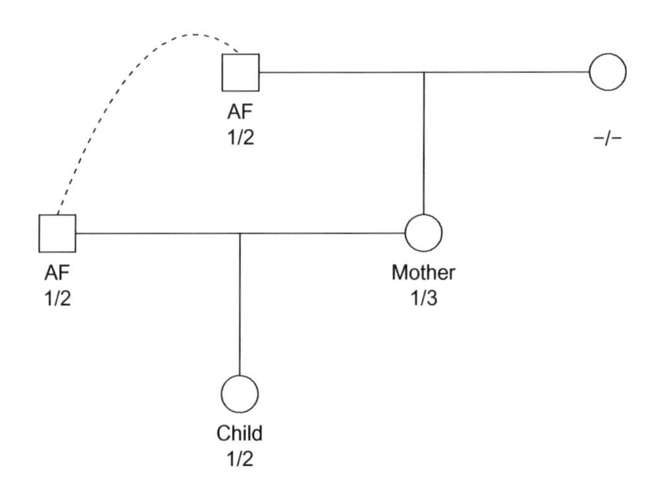

FIGURE 2.10

Incest by father. The dashed line indicates that the alleged father (AF) appears in two roles. See Exercise 2.6.

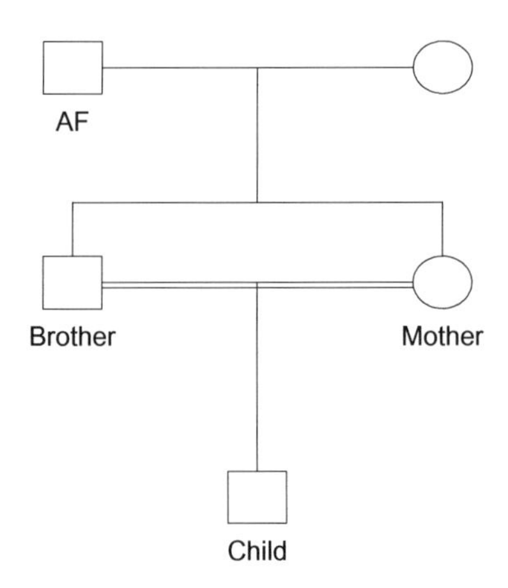

FIGURE 2.11

Incest by brother. A standard way of displaying incest is used. See Exercise 2.6. AF, alleged father.

(b) When there are more than two hypotheses, as above, some prefer to calculate posterior probabilities for the hypotheses as there is then no need to define a reference (or denominator) pedigree. Assume that each of the three hypotheses is equally likely a priori. Calculate the posterior probabilities.

Exercise 2.7 Several mutation models. In this exercise, which extends Exercise 2.2, we will try the different mutation models. Throughout we consider data from the system VWA given in Exercise 2.2. The alleged father is 14/15 and the child is 16/17, while the hypotheses are as for the duo case.

We will use the overall mutation rate $R = 0.005$ and the same model for females and males. The answer for this exercise will be obtained with `Familias`; Exercise 2.8, on the other hand, is based on theoretical calculations.

(a) Load the data `Exercise2_7.fam`. Use the equal probability mutation model. Calculate the LR. (Answer: LR $= 2.9 \times 10^{-3} = 0.0029$. *Comment*: For this model, mutations to all other alleles are equally likely).

(b) Use the `Proportional to freq.` model. Calculate the LR. (Answer: LR $= 6.3 \times 10^{-3} = 0.0063$. *Comment*: For this model, mutation to a common allele is likelier than mutation to a rare allele).

(c) Use the `Stepwise (Unstationary)` model. For this model there are two parameters. The first is as before, and should be set to $R = 0.005$. Set the second parameter, `Range`, to $r = 0.5$. Calculate the LR. (Answer: LR $= 4.7 \times 10^{-3} = 0.0047$. *Comment*: For this model, the probability of a mutation depends on the size of the mutation. With $r = 0.5$, a two-step mutation occurs with half the probability of a one-step mutation. A three-step mutation occurs with half the probability of a two-step mutation, and so on.)

(d) Use the `Stepwise (Stationary)` model with parameters as above. (Answer: LR $= 6.4 \times 10^{-3} = 0.0064$.)

(e) Use the `Extended stepwise` model with Rate 2 $= 0.1$. In this case, with no microvariants, the result is the same as for `Stepwise(Unstationary)` regardless of the value of the Rate 2 parameter.
Comment: The models `Proportional to freq.` and `Stepwise (Stationary)` are *stationary*, the others are not. If a model is stationary, introducing a new untyped person—say, the father of the alleged father—does not change the LR. This is a reasonable property of a model as introducing irrelevant information should not change the result. For an unstationary model, however, the LR will change slightly as the allele frequencies may then differ from one generation to the next. While this may appear mathematically inconsistent, it could be argued that allele frequencies change—that is, a stationary distribution has not been reached.

(f) *Verify by means of an example that the LR does not change if a father of the alleged father is introduced for the stationary models, while slight changes occur for the other models.

(g) *Comment*: There is another subtle point for all mutation models: In this case only five alleles (14, 15, 16, 17, and "rest allele") are needed rather than the eight alleles defined. A five-allele model will lead to slightly changing LRs. Verify the above and comment on it.

Exercise 2.8 *Mutation models, theoretical. In this exercise, which serves to confirm some answers obtained in Exercise 2.7, we will fill in some mathematical details related to the mutation models.

(a) Show that

$$LR = \frac{p_{16}(m_{14,17} + m_{15,17}) + p_{17}(m_{14,16} + m_{15,16})}{4p_{16}p_{17}}, \tag{2.20}$$

where p_{16} is the allele frequency for allele 16 and $m_{14,17}$ is the probability of a mutation from allele 14 to allele 17 and so on; the formula is a special case of Equation 2.12.

(b) For the equal probability mutation model $m_{ij} = m = R/(n-1)$, where n is the number of alleles, explain why Equation 2.18 in Exercise 2.2 follows from Equation 2.20. Show that when $n = 8$ and $R = 0.005$, LR $= 0.0029$, as in Exercise 2.7(b).

(c) Consider next the proportional model. By definition,

$$m_{ij} = kp_j \text{ for } i \neq j \quad \text{and} \quad m_{ii} = 1 - k(1 - p_i).$$

Show that

$$LR = k = \frac{R}{\sum_{i=1}^{n} p_i(1-p_i)},$$

and from this verify the answer in the previous exercise.

Comment: Exact calculations for the remaining mutation models are more technical. The manual for `Familias` 3 (http://familias.no/) contains some further examples.

Exercise 2.9 Paternity case with mutation. Load the file `Exercise2_9.fam`. Consider the hypotheses in Exercise 2.1.

(a) Verify that LR $= 0$.

(b) There is one marker where the child and the alleged father do not share an allele. Find this marker.

(c) Use the `Stepwise (Stationary)` model for females and males with mutation rate 0.001 and mutation range 0.5 for all markers and calculate LR.

(d) Assume you are asked to consider hypothesis H_3: The brother of the alleged father is the father. Calculate LR (H_1/H_3).

(e) Is there a "best" mutation model? Should a mutation model be used routinely for all markers?

Exercise 2.10 Sisters or half sisters? We would like to determine whether two girls (called Sister 1 and Sister 2 in Figure 2.12) are sisters (corresponding to hypothesis H_1) or if they are half sisters (corresponding to hypothesis H_2 shown on the right-hand side).

The genotypes are given in Table 2.9, with allele frequencies 0.1 for all alleles for markers S1 and S2, and 0.05 for S3-S5.

(a) What is the LR comparing the full-sister alternative with the half-sister alternative?

(b) The LR in this case does not give rise to a clear conclusion. How would you determine the required number of further markers needed for a reliable conclusion? What markers would you use? You are not asked to do specific calculations.

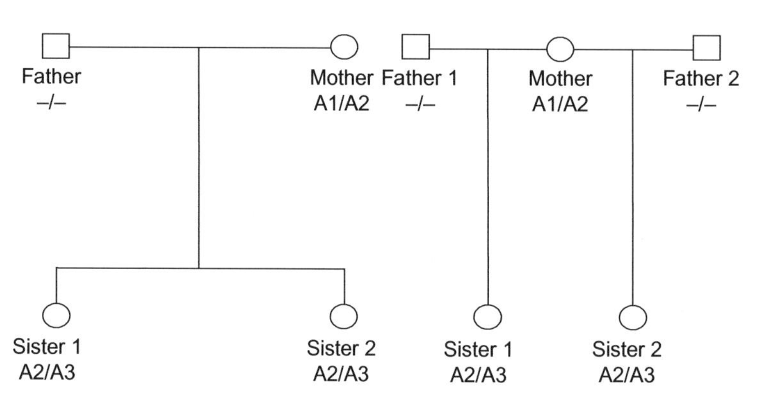

FIGURE 2.12

Pedigrees (hypotheses) in Exercise 2.10.

Table 2.9 Marker Data for Exercise 2.10

Person	S1	S2	S3	S4	S5
Mother	A1/A2	A1/A2	A2/A3	A2/A3	A2/A3
Sister 1	A2/A3	A2/A3	A3/A4	A3/A4	A3/A4
Sister 2	A2/A3	A2/A3	A1/A3	A1/A3	A1/A3

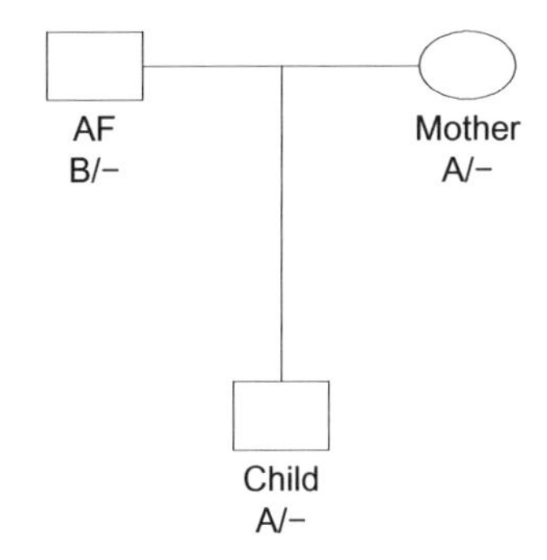

FIGURE 2.13

Pedigree for Exercises 2.11 and 2.16. AF, alleged father.

Exercise 2.11 Silent allele.

(a) See Figure 2.13. This is a paternity case where there is suspicion of a silent allele. Include a silent allele frequency of $p_S = 0.05$, and calculate the LR(father/not father) with `Familias`. The allele frequencies for A and B are $p_A = p_B = 0.1$.

(b) Confirm the above result with

$$\mathrm{LR} = \frac{p_S(p_A + p_S)}{(p_A + p_S)^2(p_B + 2p_S) + p_S p_A(p_B + 2p_S)}. \tag{2.21}$$

Exercise 2.12 *Theta correction, derivation of formula. Verify theoretically the formula in Exercise 2.1(h). *Hint*: Use the sampling formula described in Section 2.5.1.

Exercise 2.13 Theta correction, `Familias`. This exercise expands on Exercise 2.4 by introducing theta correction of 0.02. Calculate the LR based on all markers. To save time you can read the input from the file `Exercise2_4.fam`.

Exercise 2.14 Input and output. Sometimes it is of interest to read and write data, and this is the topic below:

(a) Open the file `Exercise2_3.fam`.

(b) Export the database from the `General DNA data` window. Name the output file `database.txt`.

(c) Export the case data from `Case-related DNA data`. Name the output file `casedata.txt`.

(d) Open a new project.

(e) Import `database.txt` from the `General DNA data` window.

(f) Import `casedata.txt` from the `Case-related DNA data` window.

(g) Define the pedigrees, see Exercise 2.3, and calculate the LR for all markers.

Exercise 2.15 *Paternity case with dropout.

(a) Load the file `Exercise2_15.fam`. Consider the hypotheses in Exercise 2.1. Confirm that that LR = 0, and find the one marker where the child and the alleged father do not share an allele.

(b) Set the dropout probability to 0.1 for this marker, choose `Consider dropout` in the `Case-related DNA data` window for the child, and recalculate the LR.

(c) The data could also be explained by a mutation. Compare the above result with the LR you get with the mutation model (for this marker) `Stepwise (Stationary)` for females and males with mutation rate 0.001 and mutation range 0.5. Remove the dropout. *Comment*: One could consider both dropout and mutation.

(d) Discuss: Should dropout be used for all homozygous data?

Exercise 2.16 Silent allele and dropout. Consider the paternity case in Figure 2.13. We analyze two scenarios: the first is that there is a silent allele passed on from the alleged father to the child, and the second is that there is a dropout in both the alleged father and the child. Let the allele frequencies of A and B both be 0.2.

(a) Include a silent allele frequency of 0.05 and calculate the LR with `Familias`.

(b) Remove the silent allele and instead specify a dropout probability of 0.05. We will consider dropouts for both the alleged father and the child. Calculate the LR.

Exercise 2.17 Simulation. Load the file `Exercise2_17.fam`. The hypotheses considered are as for the duo case. The file contains no genotype information. Use the simulation in `Familias` to simulate genotypes for both individuals. Untick `Random seed` and set seed to 12345. Use 1000 simulations and find the following:

(a) The mean $\text{LR}(H_1/H_2)$ when H_1 is true.

(b) The mean $\text{LR}(H_1/H_2)$ when H_2 is true.

(c) The probability of observing an LR larger than 50 when H_1 is true.

Exercise 2.18 ***A fictitious paternity case**. A child was conceived as a result of a rape. The DNA profiles of the defendant, mother, and child are available, see Table 2.10.[4] $LR_{1,2}$ of this genetic evidence for hypotheses

H_1: The defendant is the father of the child as shown in Figure 2.14.
H_2: The defendant is unrelated to the father of the child.

is very high, thus providing strong evidence for paternity of the defendant.

Now suppose that the defendant claims that he is innocent, but that he believes his brother is the actual father of the child. We formulate a third hypothesis:

H_3: The defendant's brother is the father of the child.

Table 2.10 Data for Exercise 2.18

	Locus	Mother	Child	Defendant	$LR_{1,2}$
1	CSF1PO	10/14	10/15	14/15	4.56
2	D2S1338	17/17	17/24	17/24	4.26
3	D3S1358	14/16	14/17	17/18	2.36
4	D5S818	11/13	12/13	11/12	2.83
5	D7S820	11/12	11/12	11/12	2.92
6	D8S1179	10/14	10/15	14/15	4.56
7	D13S317	8/13	12/13	12/12	3.24
8	D16S539	9/10	9/9	9/12	4.81
9	D18S51	13/14	14/18	13/18	5.45
10	D19S433	14/14	14/14	14/14	2.93
11	D21S11	29/29	29/30	30/33.2	2.15
12	FGA	22/24	24/24	22/24	3.63
13	TH01	9.3/9.3	9.3/9.3	7/9.3	1.64
14	TPOX	8/8	8/8	8/8	1.84
15	vWA	15/18	15/16	16/16	4.96
16	All				50,218,440.00

[4]We thank Klaas Slooten for this exercise. This is a fictitious case. Similar cases are sometimes encountered in practice.

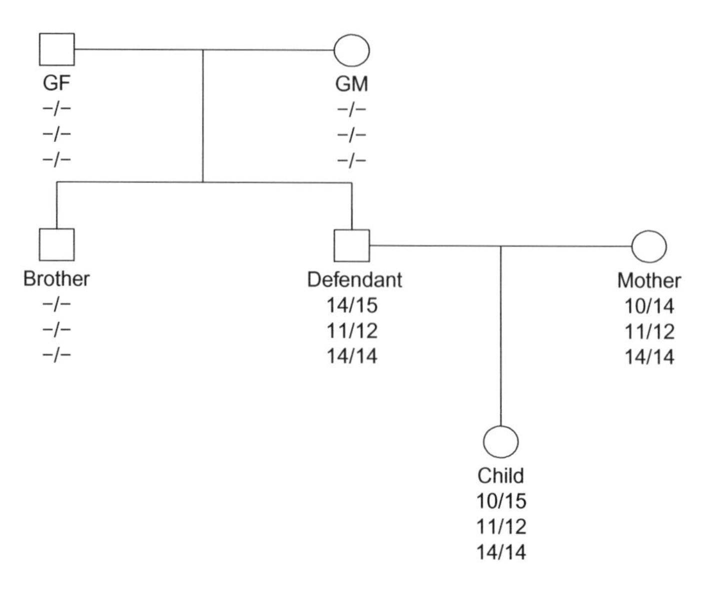

FIGURE 2.14

Hypothesis 1 in Exercise 2.18 with genotypes for CSF1PO, D7S820, and D19S433. GF, grandfather; GM, grandmother.

(a) Give the algebraic formula for $LR_{1,2}$ for loci CSF1PO, D7S820, and D19S433.
(b) Give the algebraic formula for $LR_{3,2}$ for the same loci.
(c) Can you compute $LR_{3,2}$ numerically with the information above, or do you need a table of allele frequencies? Can you explain why?
(d) What is the LR for H_1 versus H_3 based on these three loci?
(e) In the algebraic formula for $LR_{1,3}$ for locus CSF1PO, calculate its limits for $p_{15} \to 1$ and $p_{15} \to 0$, and explain the outcome.
(f) Do the same for $LR_{1,3}$ on locus D19S433 when $p_{14} \to 1$ and $p_{14} \to 0$.
(g) Discuss in the same way locus D7S820.
(h) It can be shown that $LR_{3,2} = 500$. Can you calculate the probability that each hypothesis is true?

CHAPTER

Searching for relationships 3

CHAPTER OUTLINE

This chapter introduces the concept of searching in potentially large databases for relatives. A database is defined here as a set of DNA profiles from individuals not, a priori, belonging to the same family. The search is typically an iterative procedure comparing each element of the database with some reference data and computing a probability of identity.

There is a constant increase in the number of individuals contained in criminal offender databases. For instance, in the UK alone the national database exceeds 6 million (2015),[1] and the search may not be as straightforward as in countries with smaller databases, where spurious matches are not necessarily a big concern.

[1] https://en.wikipedia.org/wiki/United_Kingdom_National_DNA_Database

Furthermore, *familial searching*, meaning that we search a database looking for relatives of a perpetrator, is a field that has attracted attention in recent years [57–60], particularly in the USA and UK owing to the success in some cold cases; see, for example, [61] for the *Surrey motorway* accident and [62] for the *Grim Sleeper* case.

Another application of databases is *disaster victim identification* (DVI). In the current setting, DVI is based on DNA data from presumed relatives of the missing persons as illustrated in Figure 1.3; see also Exercise 3.1. A particularly important subset of victim identification problems is *missing person identifications*, where databases, with increasing sizes, are kept to provide identification whenever an unidentified individual is reported. Since the successful use of DNA as a tool to identify victims of the August 1996 Spitsbergen aircraft disaster [63], numerous papers have followed, including [43, 64, 65] describing issues related to the World Trade Center 9/11 terror attack, [66] explaining problems related to the Southeast Asia tsunami in 2004, and [67] addressing the hurricane Katrina in 2005.

In all these applications, it is common to report a likelihood ratio where the underlying core computational model remains the same as for the likelihood calculations in Chapter 2.

In addition to presenting methods for DVI problems, we introduce a model for direct matching where dropouts (discussed briefly in Section 2.7), drop-ins and genotype errors are considered simultaneously.

The chapter continues the ideas presented in Chapter 2. We provide some extensions and introduce some more theoretical details. The details are outlined such that interested readers may validate and test the interface with the appropriate formulas. Readers interested only in applications can skip most of the theory and do several of the exercises as these are more or less self-contained.

3.1 INTRODUCTION

The following sections place the determination of relationships in a broader context, where we wish to find the most probable pedigrees for a large number of individuals. More specifically, there is some set of individuals (sometimes as a part of a database) that is compared with a set of reference individuals. The latter may, in turn, be composed of a set of individuals specifying a family, while it may also be a single profile. In comparison with general methods, some of the matching algorithms require swift computations to achieve results in a reasonable time. We provide some insight into algorithms and procedures for application in extended searches for relationships.

In Chapter 2, we introduced methods that are most often used to decide whether a single person X is the individual X', related to one or more other given persons, or whether X is in fact unrelated to these. Even if a standard paternity case is the most frequent example, it is also fairly common that either remains from X are found or a DNA trace from him or her is found, and the question is whether this trace represents the missing X'. The context can then either be a criminal case or a case of a missing person. We further studied how to formulate likelihoods for some genetic data given disputed relationships. This chapter will extend the formulas to include the setting where we require a likelihood for a direct match—that is, to provide a probability that two profiles originate from the same individual. A simple version of this problem was discussed in Section 2.8.1.

We generalize the argumentation to consider a list of unidentified traces $X_1, X_2, \ldots, X_n$ and a list of possible persons $X'_1, X'_2, \ldots, X'_k$ these may be matched to. Some possible contexts are as follows:

(1) Matching a single trace X_1 to a database $X'_1, \ldots, X'_k$ of criminals (traces). The trace may also be a known suspect, where the intention is to find out if he or she matches some of the elements in the database.

(2) Matching a single trace X_1 to a database with familial searching (so that each of the $X'_1, \ldots, X'_k$ are possible family members of the trace).

(3) Searching for a family's lost person X' in a database $X_1, \ldots, X_n$ of unidentified individuals, or the list of victims from a disaster.

(4) Matching a single unidentified remain X_1 to an archive $X'_1, \ldots, X'_k$ of missing people.

(5) Matching a list of traces $X_1, \ldots, X_n$ to a list of possible persons $X'_1, \ldots, X'_k$ in a DVI scenario.

(6) Searching a set of individuals $X_1, \ldots, X_n$ or $X'_1, \ldots, X'_k$ to find relations.

In each case, the intention is to find the best possible match or combination of matches given the data. The following sections deal with some theoretical points, and when describing the matching algorithms, we use the implementation in `Familias`, if relevant functionality is available [11]. In Section 2.9.3 we introduced DNA mixture evidence, and in Section 3.4.2 we will provide a statistical framework when computing likelihoods for these situations.

3.2 DISASTER VICTIM IDENTIFICATION

Disaster victim identification (DVI) indicates the collections of problems where a database of unidentified individuals, for which postmortem data (PM) is available, is compared with a set of reference data, antemortem data (AM). The latter may include a set of persons from a family where a missing person has been reported. It may also be a personal belonging of a missing person, such as toothbrush or razor. The following description and implementation may also include the application where we have a missing person database and we compare DNA from unidentified samples with DNA profiles of members of families that have reported a missing person.

When we use the phrase "missing person," we indicate a virtual person, in the sense that we may construct a pedigree where a number of reference persons have some relation to an individual, who is yet to be found. In the process that follows, as shall be illustrated, each postmortem sample is tested as the missing person.

The scale of a disaster may be contained—that is, we assume that all unidentified persons are accounted for in the list of missing persons. For instance, some reference data is available for all reported victims of a small plane crash. The opposite may not necessarily be true: all missing persons may not be accounted for among the unidentified persons. This is unavoidable following a large-scale disaster where the victims may be spread out and never found. A typical example of a contained disaster is a plane crash such as the Swissair 111 accident in 1998 [68], as previously indicated, while the South East Asia tsunami in 2004 [66] was a noncontained, open, disaster.

3.2.1 IDENTIFICATION PROCESS

Identification following a disaster commonly involves a number of laborious steps. The process typically starts with the collection of unidentified remains from the scene of the event, and forensic laboratories analyze the samples as they are received. The output data is a combination of genetic profiles as well as other characteristics, gender, etc. At the same time, the families of the victims are contacted, and all available reference data is collected. Reference data may be personal belongings of the missing person, reference samples from relatives of the missing person, and other information about the characteristics of the missing person. The step of contacting relatives and obtaining correct reference information can be cumbersome, given the uncertainty about the destiny of the missing person. It is important to collect reference information that is as accurate as possible as failure to do so can result in matches being missed. This may be particularly true if a biological relationship is misspecified, in other words an individual reports his/her missing son, while in reality it is the brother that is missing. Some measures can be taken to overcome such mistakes, as will be outlined in Section 3.2.4.

3.2.2 PRIOR INFORMATION

Following a large-scale disaster there is commonly a large amount of prior information that may subsequently be used and incorporated into a statistical model. One of the most important and useful pieces of data may be information about the gender of a missing person. Say, for instance, we have an accident with 1000 missing persons, such that the prior probability of an unidentified person belonging to a family is $1/(1000+1)$. Suppose only 10 females are reported missing. Now, given an unidentified person is a female, there are only 10 families where she may fit, given that we have a contained disaster. Thus, the prior probability is increased to $1/(10+1)$, for families reporting a missing female individual and similarly reduced to zero for families reporting a missing male individual.

We may further combine prior information with statistical evidence from a DNA comparison—that is, a likelihood calculation—with a Bayesian approach (introduced in Section 2.3.4) based on

$$\Pr(H_i \mid \text{data}) = \frac{\Pr(\text{data} \mid H_i)\Pr(H_i)}{\sum_{j=1}^{N} \Pr(\text{data} \mid H_j)\Pr(H_j)}, \tag{3.1}$$

where we would like to find the posterior probability $\Pr(H_i \mid \text{data})$ of hypothesis H_i, the probability that person i belongs to a specific reference family. The prior probability for a pedigree may be 1 divided by the total size of the accident plus 1, but other numbers may also be considered; see [69] for a brief discussion. The former prior follows from the argument that the prior probability for one of the unidentified persons belonging to a specific family is 1 divided by the number of other possible candidates plus 1 for the case where none matches the missing person. (*Note*: This assumes that all families are equally likely apriori.)

3.2.3 IMPLEMENTATION IN `FAMILIAS`

There has been an increasing demand for efficient software to handle the likelihood computations in connection with mass disasters. `Familias` [11] implements a free solution, while other commercial toolkits exist [44, 70]. Using the former, we take advantage of all the features available for case work calculations explained in Chapter 2 as well as several extensions described in this chapter. The procedure that uses the DVI module of `Familias` (version 3.0 and above) follows some elementary steps:

(i) Define a frequency database for a relevant population. For disasters involving multiple nationalities, we may specify all relevant databases and easily switch between them, allowing calculations in each of them to be performed. We define mutation parameters and silent allele frequencies. We require appropriate mutation models to be specified for all systems, and in a DVI operation this is particularly important as we may not have the possibility to investigate each match separately, but must rely solely on the LR. Also, we decide whether dropouts should be considered and specify this for each marker.

(ii) Import the set of unidentified samples, $X_1, \ldots, X_n$. This set may contain multiple remains from the same individual and may be extended at a later point as more profiles become available.

(iii) If necessary, perform a blind search (see Section 3.3). This will identify remains that are identical and in addition will provide evidence for potential relationships among the unidentified samples.

(iv) Import reference family data, $F_1, \ldots, F_m$, where we specify a missing person for each family. This is an important step which may require some extra consideration. As mentioned, each family represents one missing person with a set of reference persons. As for case work, we may need extra, untyped, persons to define the relationships. For instance, given that a missing person is reported to be the brother of the reference person, we need to define an untyped father and mother to correctly specify the sibship. Furthermore, it may be that multiple missing persons are reported for the same family, which is not directly supported in `Familias`. We may, however, use two different alternative approaches to handle this:

1. In a family, define the relation to each missing person as a separate pedigree. For instance, given that a son and a brother of the reference person are reported missing, we specify two pedigrees, one where the reference person is the father of the missing person and one where he is the brother of the missing person.
2. Specify two (or more) different reference families, with exactly the same reference persons and data in each family. Use each family to define each of the missing persons respectively.

Neither of the indicated solutions is optimal, and we will return to this with some remarks on possibly better approaches in Section 3.2.4.

(v) The last step is to perform the matching or searching. We should now obtain an extensive list with matches, which may be confirmed and reported or examined further. We may also use prior information from other sources to compute a posterior probability for each family—that is, for each family $F_1, \ldots, F_m$, we match all unidentified remains $X_1, \ldots, X_n$ and compute a posterior probability as demonstrated in Example 3.1.

`Familias` further implements a swift scan, using a zero-mutation model, to optionally be used prior to more extensive computations. The feature works by initially performing calculations with mutation rates equal to zero. The full mutation model is used only for marker likelihoods equal to zero. The number of allowed likelihoods equal to zero may be tuned. The use of complex mutation models on also comparatively simple pedigree trees may result in extensive computation times, and it is therefore recommended to allow the swift scan unless there are good arguments not to.

Example 3.1 Simple DVI case. Consider a scenario where a small sailing boat has sunk and we have retrieved remains indicating a fire. In total five remains $(R_1, R_2, R_3, R_4, R_5)$ are found. Because of severe degradation, we obtain genotype data only for a single genetic marker, M1. The genotypes are $\{12/12, 12/12, 13/14, 15/15,\ 13/13\}$ for the five remains, respectively. Three different reference families (F_1, F_2, F_3) report that they had relatives on the boat. In total, three persons are reported missing. For F_1 we have a reference father typed as 12/12, for F_2 we have a personal belonging (razor) typed as 13/14, while for F_3 we have a brother with genotype 15/15. We further use the allele frequencies $p_{12} = 0.1$, $p_{13} = 0.2$, $p_{14} = 0.05$, and $p_{15} = 0.2$.

The results from a search may be visualized as an $(N + 1) \times M$ matrix. Each row indicates an unidentified sample, with an extra line added to allow for an unknown source. Each column corresponds to a family as shown in Table 3.1. In other words, $N = 5$ is the number of unidentified remains, and $M = 3$ is the number of reference families. We compute the LR comparing each $R_1, \ldots, R_5$ against the alternative hypothesis of an unrelated individual. A flat prior is used, and so the prior probability that R_i belongs to F_1 is $1/6$, and similarly for the other families. Using a Bayesian approach, we compute the posterior probabilities for each family, such that the columns sum to 1, thereby obtaining the most probable match for each

Table 3.1 Posterior Probabilities for the Reference Families (F_1, F_2, F_3) and the Unidentified Remains ($R_1, \ldots, R_5$)

	F_1	F_2	F_3
R_1	0.476	0	0.0227
R_2	0.476	0	0.0227
R_3	0	0.98	0.0227
R_4	0	0	0.818
R_5	0	0	0.0227
?	0.048	0.02	0.0912

family. The results suggest that R_1 and R_2 belong to F_1 (and the data even suggest that these two may be from the same victim), while R_3 may belong to F_2. For R_4 we have a good match as the missing brother in F_3, while for the last remain, R_5, there is no good match (detailed calculations and assumptions are omitted). R_5 may have the same origin as the profile of R_3 but due to the low quality of the samples, the genotypes may contains errors (we return to this topic later). An important observation from Table 3.1 is that the probabilities sum to 1 for each column—that is, for each reference family as pointed out above. As a consequence, one unidentified remain may have a total posterior probability above 1 when a row is summed; see, for instance, R_3.

3.2.4 EXTENSIONS

Quick searching

`Familias` implements further functionality to perform quick searches. As for blind search, we are interested in an algorithm that blindly compares the postmortem samples with the antemortem samples without prior knowledge of the relationships. In other words, omitting the information about pedigree structures as defined for each family, we may utilize a rapid search conducting pairwise comparisons where we take each unidentified sample and match it against all reference persons as exemplified below.

Example 3.2 Quick search DVI. Consider data from two parents of a missing daughter. The genotypes for a single genetic marker are 14/14 for the father and 17/18 for the mother. An unidentified person is genotyped as 18/23. We compute the LR for the match, where the unidentified person belongs to the family, and we get (disregarding the event of double mutations and other complications)

$$\begin{aligned}
\mathrm{LR} &= \frac{\Pr(\text{data} \mid \text{unidentified person belongs to the family})}{\Pr(\text{data} \mid \text{unidentified person is unrelated to the family})} \\
&\approx \frac{\Pr(14/14)\Pr(17/18) \times 0.5 \times m_{14,23}}{\Pr(14/14)\Pr(17/18)\Pr(18/23)} \qquad (3.2) \\
&= \frac{m_{14,23}}{4p_{18}p_{23}},
\end{aligned}$$

where $m_{14,23}$ denotes the probability of a mutation from allele 14 to allele 23, which is extremely low for a stepwise model. In fact, it is very likely that $m_{14,23} \ll 4p_{18}p_{23}$, even if alleles 18 and 23 are uncommon. Thus, the evidence points against the unidentified person belonging to the family. What if we disregard the data from the father? We would then have

$$\begin{aligned} \text{LR} &= \frac{\Pr(\text{data} \mid \text{unidentified person and mother are related as parent/child})}{\Pr(\text{data} \mid \text{unidentified person and mother are unrelated})} \\ &= \frac{\Pr(17/18) \times 0.5 \times p_{23}}{\Pr(17/18)\Pr(18/23)} \\ &= \frac{1}{4p_{18}}, \end{aligned} \tag{3.3}$$

which, given that the frequency p_{18} is lower than 0.25, provides evidence in favor of the relationship. The point is to illustrate the importance of not always trusting the stated relationship; an alleged father may, unknowingly, not be the father. Even though we would assume that the mother should be aware of this, she may not want to share this information or may even be unaware of the consequences for the identification. A blind search (quick search) is always recommended as a first step to detect relationships that may otherwise be overlooked. Even though this quick search may not always be used to provide final conclusions, we can use it as a pointer when performing the real matching.

Multiple relatives

A situation which may not be uncommon is the event where a reference family indicates two (or more) missing persons. In other words, some of the reference families $F_1, \ldots, F_m$ may contain two or more of the missing persons $X'_1, \ldots, X'_k$. As introduced in Section 3.2.3, ad hoc solutions exist in `Familias` where we, for instance, may specify different pedigrees for the same family, with one missing person in each pedigree. This may, however, lead to obstacles as we can have situations where there is a dependency between the missing persons. For instance, consider the example where we have reference data from a female, whose daughter is missing, while in addition the father of the daughter, also the full brother of the mother, is reported as missing. The data for the two missing persons are obviously not independent, which will bias the calculations.

Another example where the approach considered above may be insufficient, is when we have reference data from a male, reporting a missing son and spouse. The spouse cannot be identified with the information from the husband alone. However, using the information that the missing son is fitted as the child of the male, one may subsequently identify the mother using this information.

The most general solution to the problem would take into consideration and test all possible pedigree setups with multiple missing persons in the same pedigree. In other words, if a family reports one missing person, N different comparisons are needed, where N is the number of unidentified samples. If a family reports two missing persons, we need $N \times (N - 1)$ comparisons. If the argumentation is extended to a family reporting $J < N$ missing persons, the number of comparisons needed grows to $N \times (N - 1) \cdots (N - J + 1)$. The approach may be viewed as a brute force method, where all possibilities are considered. Given large N and J, the number of computations grows fast, and each calculation may in addition be computer intensive. To illustrate, in a search where $N = 1000$ and the number of families $M = 200$ and for each family $J = 2$, the total number of computations would be $200 \times 100 \times 999 \approx 2 \times 10^8$, which is, in cases of complicated family structures, computationally too demanding.

Another, computationally appealing, approach is to perform a clustering. We start with clustering the unidentified samples, $X_1, \ldots, X_n$, using a *blind search*, as described in Section 3.3. This is done pairwise and we should remember to keep the information from the clustering in the further matching against the X_k''s. If two samples with some differences have been merged, there is larger uncertainty in the resulting profile.

We continue by calculating the LR for a single missing person X_j', in each family, thus obtaining a list of LRs indicating the most probable family for each unidentified profile. This would be followed by a second search for all families indicating more than one missing person. Consider, for example, the case with a reference family indicating two missing persons, $j = 1, 2$. The first search would produce a list with relevant matches for each of the two j's. We can now choose different approaches to continue. One solution is to test all combinations of elements of the two lists. Given that the lists are fairly small (compared with the complete lists of unidentified samples), the number of computations is acceptable. We would obtain a second list with joint LRs for different combinations of elements from the two lists. It would be necessary to provide the individual lists and the joints lists in the results. In comparison, consider that we decide to continue with the top 10 candidates from the first search. Again $N = 1000$, $M = 200$, and for each family, $J = 2$. The total number of computations would then be $200 \times 1000 \times 10^2 = 2 \times 10^7$, which may be a more acceptable number of computations.

The above approach may, however, be unsuccessful as matches may still be missed. For instance, it may be that the mutual information given by the combination of the profiles from two unidentified persons provides sufficient evidence for identification, while the single information from each of them is insufficient. Using an extension of the above approach, we may continue the search for all families indicating no matches and apply the brute search method indicated above.

Another complication, arising when matching multiple missing persons into one family, is how to weigh the likelihoods. For instance, consider the case where a father reports a missing son and a missing brother. In the first search we compare hypotheses

H_1: The father is the parent of the X_n under consideration.
H_2: The father is unrelated to the X_n under consideration.

for each unidentified remain X_n and compute a list of LRs for all N unidentified remains. The next step would compare hypotheses:

H_1: The father is the brother of the X_n under consideration.
H_2: The father is unrelated to the X_n under consideration.

and similarly compute the LR for all unidentified remains. We now have two subset lists, $X^{(1)}$ containing the list of matches from the first search and $X^{(2)}$ containing the list of matches from the second search. In the final step it would be necessary to compare hypotheses:

H_1: The father is the parent of the $X_i^{(1)}$ and the brother of the $X_j^{(2)}$ under consideration.
H_2: The father is unrelated to the $X_i^{(1)}$ and the brother of the $X_j^{(2)}$ under consideration.
H_3: The father is the parent of the $X_i^{(1)}$ and unrelated to the $X_j^{(2)}$ under consideration.
H_4: The father is unrelated to the $X_i^{(1)}$ and the $X_j^{(2)}$ under consideration.

The number of different hypotheses we need to consider grows with the number of missing persons for each family. In the current example, it would perhaps be natural to scale the LRs versus H_4, but this needs to be clear when creating the expert report for identification. In addition, careful consideration is needed by the forensic scientist. If we, for instance, report a high $\text{LR}_{1,4}$, comparing the evidence under H_1 and H_4, it may still be that $\text{LR}_{2,4}$ is even greater, and thus a false match may be reported.

3.3 BLIND SEARCH

Several situations may require a blind search in a list of individuals. For instance, in an accident the remains of the victims may be scattered such that we have multiple profiles from the same individual. In addition, the victims may be related and we may wish to use the blind search to cluster profiles that may be connected. Furthermore, when we collect data for frequency databases, it is crucial that the individuals are not closely related. Using the blind search, we may sift out related profiles, leaving us with a set of unrelated individuals. Moreover, we may search any set of persons trying to find relations that may exist. We divide the search into two different categories, *kinship matching* and *direct matching*, and describe them in detail below.

3.3.1 KINSHIP MATCHING

A blind search requires a fast procedure for the computations as we may potentially have a large number of comparisons. Consider, for instance, a set of 1000 profiles genotyped for 17 genetic markers. The total number of computations (LRs calculated) where we compare all against all reaches $999 \times 500 \times 17 = 8{,}491{,}500$. It suffices to say that given cumbersome calculations for each individual marker, the computation time will be unacceptable.

As the blind search interface uses predefined relationships, we may use the identity by descent (IBD) formula, as introduced in Section 2.3.3, for pairs of individuals. We specify the following:

H_1: Two profiles, G_1 and G_2, are related as R.
H_2: G_1 and G_2 belong to two unrelated individuals.

In the current implementation, $R \in$ {Parent–Child, Siblings, Half-siblings, Cousins, Second Cousins}.

To briefly recapitulate Equation 2.6, we define the probabilities $\Pr(\text{IBD} = i \mid R) = k_i$ and $\Pr(G_1, G_2 \mid \text{IBD} = i)$, where i denotes the number of alleles IBD. Exercise 6.5 demonstrates how IBD coefficients are calculated and also includes the inbred case.

Ordinary calculations in `Familias` use the Elston-Stewart algorithm [31], thus summing over all possible genotypes for the founders. Using instead the IBD approach, we have to perform only a single line of calculation. Even though this includes several multiplications, the computations will be much swifter. The formula is in, addition, easy to adjust for subpopulation effects.

For parent–child relations we need another approach as we require models for both mutations and silent alleles. A general formula was introduced in Equation 2.12, where mutation was considered for a duo. To allow for silent alleles, an extension is required. The exact formulation is omitted here, see Example 6.8 for details.

3.3.2 DIRECT MATCHING

In forensic genetics applied to criminal case work, it is common to report the random match probability (RMP), which may be interpreted as the probability that a random individual, unrelated to the person of interest, will have the same profile as the person of interest. The inverse of the RMP (1/RMP) is a version of the LR. See also Section 2.8.

In the current context, it may be necessary to provide a value indicating whether two profiles come from the same individual. We use a general notation and specify the following:

H_1: Two profiles, G_1 and G_2, belong to the same individual.
H_2: G_1 and G_2 belong to two unrelated individuals.

We will consider a general model for direct matching as introduced by Kling et al. [11]. The model may be written as

$$\begin{aligned}\text{LR} &= \frac{\Pr(G_1, G_2 \mid H_1)}{\Pr(G_1, G_2 \mid H_2)} \\ &= \frac{\sum_{i=1}^{N} \Pr(G_{\text{true},i})\Pr(G_1 \mid G_{\text{true},i})\Pr(G_2 \mid G_{\text{true},i})}{\Pr(G_1 \mid H_2)\Pr(G_2 \mid H_2, G_1)},\end{aligned} \tag{3.4}$$

where in the numerator we sum over all the possible genotypes for a latent genotype, denoted $G_{\text{true},i}$. The latent genotype is considered here to be the *true* genotype, unknown before the calculations. Given H_1, we propose that the two observed genotypes both originate from the latent genotype. Given the general model, this may be true, even though $G_1 \neq G_2$ as there may be genotyping errors. For the denominator, the probabilities $\Pr(G_1 \mid H_2)$ and $\Pr(G_2 \mid H_2, G_1)$ are also summations over the possible true genotypes for the individuals. However, for computational reasons, the observed genotypes are assumed to be fixed, and we have $\Pr(G_1 \mid H_2) = \Pr(G_1)$ and $\Pr(G_2 \mid H_2, G_1) = \Pr(G_2 \mid G_1)$.

The probabilities in the summations are explained as follows: $\Pr(G_{\text{true},i})$ is the probability of the latent genotype, given as a product of allele frequencies, where θ correction may be applied. Further, $\Pr(G_1 \mid G_{\text{true},i})$ and $\Pr(G_2 \mid G_{\text{true},i})$ are the conditional probabilities for observing genotypes G_1 and G_2 given the latent genotype. These are given as functions of the probabilities for observing a dropout (d), a drop-in (c), and a typing error (e). Recall from Chapter 2 that dropout is defined as the event when a homozygous profile is observed even though the true genotype is heterozygous—that is, one allele has dropped out. Several models for dropout are discussed for kinship cases in Dørum et al. [46]. Drop-in is defined as the event when an extra allele is observed—for example, a heterozygous profile is observed even though the true genotype is homozygous. The last parameter has a looser definition as any error being caused in the laboratory as when a wrong genotype is assigned for technical reasons. Here we consider errors on the level of the genotype whereas in Example 6.10, page 175, we consider errors on the level of the individual alleles.

We will see an example of the effect of these parameters in Example 3.3. We note that if d, c, and e are zero, the conditional probabilities will be zero if $G_1 \neq G_2$ and 1 otherwise. Given that all parameters are zero and $G_1 = G_2$,

$$\begin{aligned}\text{LR} &= \frac{\Pr(G_1)}{\Pr(G_1)\Pr(G_2 \mid G_1)} \\ &= \frac{\Pr(G_1)}{\Pr(G_1)\Pr(G_2)} = \frac{1}{\Pr(G_2)},\end{aligned} \tag{3.5}$$

where the second equality holds if $\theta = 0$. The rightmost part is recognized as 1/RMP.

Example 3.3 Direct matching algorithm. We will consider a single nucleotide polymorphism marker with alleles A/a and corresponding frequencies $p_A = 0.2$, $p_a = 0.8$. Furthermore, assume we have two profiles, one from a crime stain (G_1), genotyped as A/A, and the other from another stain from the same crime scene (G_2) with genotype A/a. We suspect they may be from the same perpetrator and we wish to compute the LR for this. Using Equation 3.4, we get (if $\theta = 0$)

$$\begin{aligned}
\text{LR} &= \frac{\Pr(G_1, G_2 \mid H_1)}{\Pr(G_1, G_2 \mid H_2)} \\
&= \frac{\sum_{i=1}^{N} \Pr(G_{\text{true},i})\Pr(G_1 \mid G_{\text{true},i})\Pr(G_2 \mid G_{\text{true},i})}{\Pr(G_1)\Pr(G_2)} \\
&= \frac{\begin{array}{l}\Pr(\text{A/A})\Pr(G_1 \mid G_{\text{true},i} = \text{A/A})\Pr(G_2 \mid G_{true,i} = \text{A/A}) \\ + \Pr(\text{A/a})\Pr(G_1 \mid G_{\text{true},i} = \text{A/a})\Pr(G_2 \mid G_{\text{true},i} = \text{A/a}) \\ + \Pr(\text{a/a})\Pr(G_1 \mid G_{\text{true},i} = \text{a/a})\Pr(G_2 \mid G_{\text{true},i} = \text{a/a})\end{array}}{\Pr(\text{A/A})\Pr(\text{A/a})}.
\end{aligned}$$

Consider the first term of the sum in the numerator. Pr(A/A) is easy to calculate, while the other two terms require some elaboration. $\Pr(G_1 \mid G_{\text{true},i}{=}\text{A/A})$ evaluates to a function corresponding to $(1-d^2)(1-e)(1-c)$, where specific derivations are omitted. Briefly, the probability is a product of the probability of not observing a dropout, given by $2(1-d)d + (1-d)^2 = 1-d^2$, and the probability of not observing a drop-in, given by $(1-c)$, and the probability of not observing a typing error, given by $(1-e)$. All events are assumed to occur independently. Similarly, $\Pr(G_2 \mid G_{\text{true},i}{=}\text{A/A})$ evaluates to a function corresponding to $e(1-c)(1-d^2) + (1-e)(1-d^2)cp_a$, where we in the first part consider that G_2 is produced through a genotyping error and the last part corresponds to the event that a drop-in of allele a has occurred. Using similar reasoning for the rest of the terms, we get for $d = 0.1$, $c = 0.01, e = 0.001$, and

$$\begin{aligned}
\text{LR} &= \frac{\Pr(G_1, G_2 \mid H_1)}{\Pr(G_1, G_2 \mid H_2)} \\
&= \frac{\begin{array}{l} p_A^2(1-d^2)(1-e)(1-c) \times \left(e(1-c)(1-d^2) + (1-e)(1-d^2)cp_a\right) \\ + 2p_Ap_a \times \left(e(1-c)(1-d)^2 + (1-e)(1-d)d(1-c)\right) \times (1-d)^2(1-e)(1-c) \\ + p_a^2 \times e(1-d^2)(1-c) \times \left(e(1-d^2)(1-c) + (1-e)(1-d^2)cp_A\right)\end{array}}{2p_A^3p_a} \\
&= \frac{\begin{array}{l} 0.2^2 \times 0.979 \times (0.00098 + 0.0079) \\ + 2 \times 0.2 \times 0.8 \times (0.0008 + 0.089) \times 0.8 \\ + 0.8^2 \times 0.00098 \times (0.0008 + 0.00198)\end{array}}{2 \times 0.2^3 \times 0.8} \approx 1.8,
\end{aligned}$$

which can be confirmed in `Familias`. This indicates that even though the profiles are different, the evidence suggests that the profiles may be from the same individual. *Note*: Some simplifying assumptions are necessary—for instance we disregard the term corresponding to two alleles dropping out and two others dropping in.

3.4 FAMILIAL SEARCHING

Stains from crime scenes are regularly searched for direct matches in large national or local database of convicted offenders. The opposite may also be true: a suspect that is matched up against a database of stains. It is likely that a combination of the two methods is applied. We may report an LR using methods as described in Section 3.3.2 or by merely stating the number of matching markers. Ideally two profiles should have exactly the same genotypes at all markers in order for a match to be reported. However, to allow for dropouts, a perfect match is frequently not required, as discussed in [71]. In contrast to direct matching, *familial searching* amounts to searching for relatives of a stain in the database, a quest for relatives of the true perpetrator. This may typically be carried out to follow up on a direct matching resulting in no leads, but a lower-stringency search indicates some potential matches. As mentioned above, we may also investigate if a suspect is the brother of an individual for some stain in a database. Familial searching is a fairly new concept, but the basic ideas are the same as those discussed in Chapter 2, and the underlying concept is merely an extension of applying existing methods to a larger search. An introduction is provided in Bieber et al. [57], and several other authors have discussed the underlying calculations [58–60, 72].

Familial searching is not allowed in some countries, while in others the legal situation is unclear. These issues will not be discussed here; we merely note that there is concern about personal integrity. See, for example, Suter et al. [73] for a discussion of some of the ethical aspects. There have, however, been a number of reported successful uses in the literature, with one of the most famous being the *Grim Sleeper* case [62]. To briefly recapitulate, a serial killer was on the loose in Los Angeles, USA, connected to at least 10 different murders in the 1990s. No direct matches were ever found in the DNA database, though following a familial search in the 2009, a parent–child match was produced. The findings suggested that a parent or child of the offender in the database had committed the crimes. A suspect, the father of the convicted offender, was apprehended in 2010, and a perfect DNA match was confirmed.

Example 3.4 Familial searching—a simple case. Imagine a database of four elements (E_1, E_2, E_3, E_4) with single locus profiles (12/12, 13/14, 14/15, 15/15). The corresponding allele frequencies are given by $p_{12} = 0.1, p_{13} = 0.2, p_{14} = 0.2$, and $p_{15} = 0.1$. A trace is found at a crime scene with the genetic profile 12/13. A direct match search is performed indicating no perfect matches, and the investigators decide to try a familial search instead. We compare the trace with all database elements looking for parent–child and sibling relations. For each comparison we compute an LR as illustrated in Table 3.2. Using this information, the investigators may decide to go further and genotype more than this single marker! In other circumstances, a typical procedure would be to genotype Y-chromosomal markers (given that the trace is of male origin) for a certain number of the matches. These markers may subsequently be used to exclude false matches, further reducing the list of matches. *Note*: The calculations disregard complications such as mutations and silent alleles.

Table 3.2 LRs Comparing the Trace with the Database Elements (E_1, E_2, E_3, E_4)

	LR (Parent–Child)	LR (Siblings)
E_1	5	2.75
E_2	1.2	0.875
E_3	0	0.25
E_4	0	0.25

3.4.1 IMPLEMENTATION

`Familias` implements a version of familial searching. Similarly to what is done in normal case work, we define the frequency database containing the data for the population under consideration. We may actually have several different population frequency databases in the same project and can easily switch between them. Similarly to the DVI interface, described in Section 3.2.3, the familial searching procedure may be outlined from some basic steps:

(i) *Database*. First, we start by importing the data set we wish to perform the search on, usually a larger database contained in a software program such as `CODIS`, developed by the FBI. As described on the homepage,[2] `CODIS` is not directly built for familial searching, even though some simple means are provided to search for allele sharing. In `Familias`, the database may contain single-source profiles or mixtures. In fact, simple mixtures can be imported and searched for; see Section 3.4.2.

(ii) *Blind search* (optional). Perform a blind search on the database. This is most likely not necessary, but is useful if the investigator wishes to cluster profiles in the database. Duplicate entries can be resolved.

(iii) *Profiles*. We continue with defining a set of profiles, stains/suspects/etc., to search for. These profiles may be single-source profiles or mixtures and may have complete or partial overlap with the set of markers each database element is genotyped for.

(iv) *Search options*. We further define search parameters:

- **(1)** `LR threshold`, used to put some constraints on the number of matches.
- **(2)** `Relationship`, a list of relationships or a direct match.
- **(3)** `Scale versus`, used to select the alternative hypotheses. In normal circumstances unrelated is chosen, but may in some situations be specified otherwise—for example, if the population under consideration is known to contain related individuals.
- **(4)** θ, subpopulation correction (see Section 2.5).

[2]See http://www.fbi.gov/about-us/lab/biometric-analysis/codis.

(5) `Dropout`, `Dropin`, and `Typing error` probabilities are specified for direct matching only. See Section 3.3.2 for details.

In addition, for parent–child relations mutation parameters and silent alleles are specified for each marker.

(v) *Matching*. The last step is to perform the search. Calculations are performed for all overlapping markers, presenting an exhaustive list with the results. The matching is performed pairing each database element with each profile or trace.

The result from the familial searching is a list of LRs for each match. We may also compute posteriors, however the definition of prior probabilities is not straightforward. We may use 1 divided by the size of the database. This will, however, in many cases result in a high number of errors; see Example 3.5. We therefore recommend the LR, which does not require a prior, and rather use other measures to reduce the number of possible matches—for instance, by typing Y-chromosomal markers.

Example 3.5 Familial searching and use of *Bayes's* theorem. Consider a frequency database for 16 standard short tandem repeat (STR) markers. Assume we have a database of convicted offenders of size N, genotyped for those 16 markers. Furthermore, we are interested in searching for a sibling of a profile found at a crime scene, but are uncertain about what LR threshold to use. Using simulations, we find the approximate distribution of the LR for a pair of siblings (see Figure 3.1).

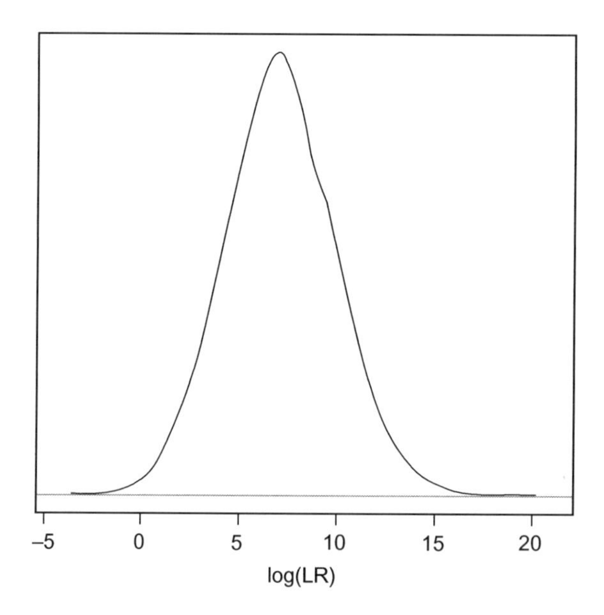

FIGURE 3.1

Distribution of log LR (by default log denotes the natural logarithm in this book) for a case of siblings with unrelated as the alternative hypothesis. A standard set of 16 STR markers has been used. The log LR is approximately normally distributed with mean 7.1 and standard deviation 2.8.

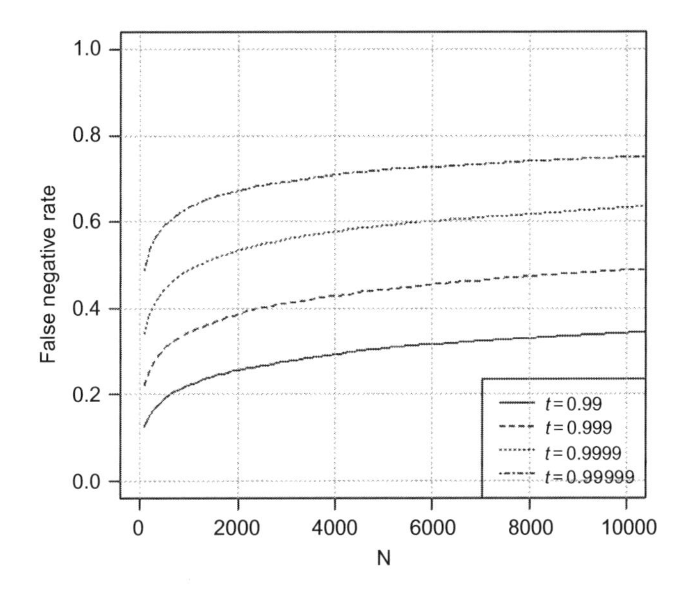

FIGURE 3.2

False negative rate versus the size of the database (N).

We may instead compute the posterior probability for each simulated case using *Bayes'* theorem and a prior of $1/N$, corresponding to the size of the database. In Figure 3.2 the false negative rate is plotted versus N for some values on the threshold for inclusion (t). For the standard set of 16 STR markers used in this example, we see that the false negative rate increases rapidly to begin with and then slowly levels off. We also note that, for instance, if $t = 0.9999$, the false negative rate is well above 0.8 for a database of 100,000 individuals, indicating that we would miss 80% of the true siblings in a search. In other words, if the database is large, the prior probability will conversely be low and as a consequence a higher LR will be required to exceed the posterior probability threshold.

We are here primarily interested in the number of false negatives as these are the true matches we miss when applying a certain prior probability. The false positives may be addressed similarly (see also Section 3.4.3), and we may address the cost of including too many false candidates (see Section 8.1). Certainly, the latter may be interesting as we at this point include rather too many candidates in order to rule them out later using other means, while the cost of this should not be ignored.

3.4.2 *RELATIVES AND MIXTURES

In databases with stains and traces from crime scenes, mixture evidence is not uncommon. Mixtures, introduced in Section 2.9.3, are profiles originating from more than one contributor. This does not impose restrictions on the minimum number of observed alleles for each genotype, but rather suggests that we may observe more than two different alleles for a marker. For instance, the genotype observed as 12/12 may in fact be a mixture of several contributors, all with genotypes 12/12. In comparison, the genotype observed as 12/13/14 is a mixture of at least two contributors (disregarding events such as trisomies and drop-ins). We here introduce some basic ideas in order to compute the likelihood that an individual is the relative of some contributor to a mixture. For further discussion of mixtures and relatives, see [74, 75] for an introduction and Chung et al. [76, 77] for applications in a database searching context.

Example 3.6 Likelihood computations for mixtures. We start by considering a simple example where one profile, S_1, is a mixture, assumed to consist of two contributors, while a second profile S_2 is a single-source profile. We may define the following:

H_1: S_1 is a mixture of S_2 and an unrelated individual, u_1.
H_2: S_1 is a mixture of two unrelated profiles (u_1 and u_2), while S_2 is unrelated to the profiles in the mixture.

Using condensed notation and disregarding observation level errors, we may now write the LR as

$$\mathrm{LR} = \frac{\Pr(S_1, S_2 \mid H_1)}{\Pr(S_1, S_2 \mid H_2)} = \frac{\sum_{u_1} \Pr(u_1, S_2 \mid H_1)}{\Pr(S_2) \sum_{u_1,u_2} \Pr(u_1, u_2 \mid H_2)},$$

where the sum in the numerator extends over all profiles such that the union of u_1 and S_2 is S_1. For the denominator, the union of u_1 and u_2 is S_1. Furthermore, in a familial searching context, where we search for relatives, we use an alternative formulation:

H_1: S_1 is a mixture of a relative of S_2, denoted R_1, and an unrelated individual U_1.
H_2: Remains unchanged.

We formulate the LR as

$$\mathrm{LR} = \frac{\Pr(S_1, S_2 \mid H_1)}{\Pr(S_1, S_2 \mid H_2)} = \frac{\sum_{r_1,u_1} \Pr(u_1, S_2, r_1 \mid H_1)}{\Pr(S_2) \sum_{u_1,u_2} \Pr(u_1, u_2 \mid H_2)}, \tag{3.6}$$

where again the summations are consistent with the mixture. To illustrate this, consider some genetic marker data where $S_1 = 12/13$ and $S_2 = 12/12$. We wish to compute the LR given that a brother of S_2 may be in the mixture S_1. Even though S_1 may appear to be a single-source profile, other evidence could suggest otherwise—for example, peak heights, and information from other markers. Given H_1, we may list all possible joint values R_1 and U_1 can take (see Table 3.3). We now have all the ingredients to complete the sum in the numerator of Equation 3.6. A similar table can be constructed for the denominator. Using $p_{12} = 0.1$ and $p_{13} = 0.2$, we get

Table 3.3 Possible Genotypes for R_1 and U_1 given a Mixture $S_1 = 12/13$ and a Profile $S_2 = 12/12$

R_1	U_1	Pr(U_1)	Pr(R_1, S_2)
12/12	12/13	$2p_{12}p_{13}$	$0.25p_{12}^4 + 0.5p_{12}^3 + 0.25p_{12}^2$
12/12	13/13	p_{13}^2	$0.25p_{12}^4 + 0.5p_{12}^3 + 0.25p_{12}^2$
12/13	12/12	p_{12}^2	$0.25 \times 2p_{12}^3p_{13} + 0.5p_{12}^2p_{13}$
12/13	12/13	$2p_{12}p_{13}$	$0.25 \times 2p_{12}^3p_{13} + 0.5p_{12}^2p_{13}$
12/13	13/13	p_{13}^2	$0.25 \times 2p_{12}^3p_{13} + 0.5p_{12}^2p_{13}$
13/13	12/12	p_{12}^2	$0.25p_{13}^2p_{12}^2$
13/13	12/13	$2p_{12}p_{13}$	$0.25p_{13}^2p_{12}^2$

$$\mathrm{LR} = \frac{2p_{12}p_{13} \times 1 \times \left(0.25p_{12}^4 + 0.5p_{12}^3 + 0.25p_{12}^2\right) + \cdots + 2p_{12}p_{13} \times 1 \times \left(0.25p_{12}^2p_{13}^2\right)}{p_{12}^2\left(p_{12}^2 \times 2p_{12}p_{13} + \cdots + p_{13}^2 \times 2p_{12}p_{13}\right)}$$

$$= \frac{\begin{array}{c}2 \times 0.1 \times 0.2 \times \left(0.25 \times 0.1^4 + 0.5 \times 0.1^3 + 0.25 \times 0.1^2\right) \\ + \cdots + 2 \times 0.1 \times 0.2 \times \left(0.25 \times 0.1^2 \times 0.2^2\right)\end{array}}{0.1^2\left(2 \times 0.1^3 \times 0.2 + \cdots + 0.2^2 \times 2 \times 0.1 \times 0.2\right)} \approx 5.4.$$

From Example 3.6 we see that it is necessary to condition on the number of contributors, which in many cases is unknown. There are methods to estimate the number of contributors to a mixture; see Egeland et al. [78]. The implementation in `Familias` combines the possibilities that either two or three individuals contributed to the mixture—that is, we compute two likelihoods, the first assuming two contributors and the second assuming three. Conditioning on more contributors increases the computational effort considerably and has therefore been omitted.

The formulas presented above assume no observation-level errors, but the equations may be easily extended to include concepts such as dropouts. The obstacle lies in the fact that this introduces another uncertainty about the genotypes and therefore another summation, consequently leading to a large number of computations.

Example 3.7 Relatives in mixtures. We may consider an extension of the ideas presented in Example 3.6, where the contributors to a mixture are in addition related. For instance, we may have the following:

H_1: S_1 is a mixture of a mother, M, and her child C. S_2 is the father of the child.
H_2: S_1 is a mixture of M and C, while S_2 is unrelated to the profiles in the mixture.

Using the same frequency data as in Example 3.6, we get (omitting derivations) LR = 4.444. In the line of reasoning above, we assumed the mother is unavailable for genotyping.

Using the same frequency data as above and assuming the mother is genotyped as $M = 12/13$, we get (omitting derivations) LR = 3.3333, which illustrates the importance of typing as many contributors as possible in a mixture to reduce the uncertainty. In the current case, the LR is slightly lowered given the genotype of the mother is known.

3.4.3 SELECT SUBSETS

Whenever a familial search is performed, we can expect a number of spurious or adventitious matches. This is particularly true the more distant the relationship we search for is. For instance, when we are searching for parent–child relationships, disregarding mutations, both individuals must share one allele at each marker, and consequently the number of matches may be low. Disregarding mutations is, however, unwise, and spurious matches can be expected even for a parent–child search. Searching instead for a half sibling of the perpetrator, we can expect several spurious matches (see Example 3.8) as there is a 0.5 probability that two individuals share zero alleles at a marker given the indicated relationship.

Example 3.8 Searching for a half sibling. Consider a database of size N, where all individuals are genotyped for 16 autosomal STR markers. Moreover, consider a stain found at a crime scene (genotyped for the same 16 markers) which is to be searched against the database looking for a half sibling. We perform simulations in `Familias` to demonstrate the overlap between the LR for half siblings and unrelated individuals (see Figure 3.3). The figure illustrates a big overlap between

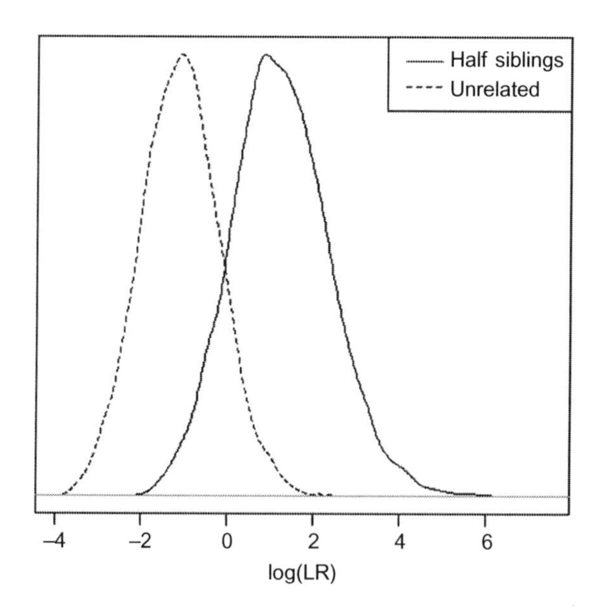

FIGURE 3.3

The log LR distribution for a pair of half siblings (to the right) and unrelated individuals (to the left) overlap.

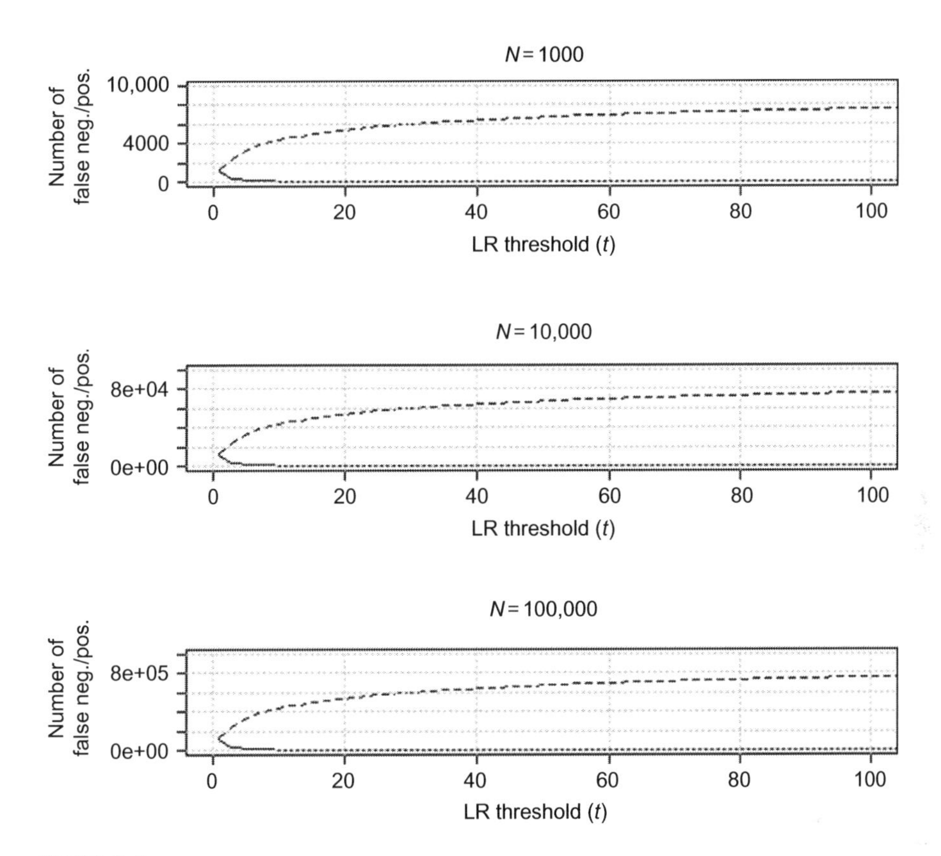

FIGURE 3.4

The number of false positives and false negatives in a database of different sizes (N) obtained with a threshold for inclusion of t. The filled lines indicate false positives, while the dashed lines indicate false negatives.

the approximate distribution under the half-sibling hypothesis and the distribution under the unrelated hypothesis.

Using the information in Figure 3.3, we may now further plot the number of expected false negatives/false positives, against the threshold for including a match for further investigation for a number of different values on N. From Figure 3.4 we see that there is no perfect threshold, rather we must make a trade-off between the costs of false inclusion versus false exclusion. In other words, when the LR threshold is increased, the number of false negatives (exclusions) increases while the number of false positives (inclusions) decreases. For instance, taking a closer look at the graph illustrating $N = 1{,}000$, we see that the number of false negatives, i.e., matches that will be overlooked, approaches 8,000 for $t = 100$.

The point is to illustrate that we need further refinements to constrain the number of matches to pursue. Kruijver et al. [79] demonstrated four different subsetting strategies—that is, methods to select subsets of a search to explore, and their performance. These are briefly described below (Familias implements versions of the three first subsetting methods.).

Top k

The *top-k* strategy is a naive approach where we select only the first k number of matches in a search. The approach disregards the evidence values. Several of the matches not included among the top k matches may have comparatively high LR. The reason for choosing this strategy may be that it is convenient to proceed with a certain number of candidates each time.

LR threshold

Another simple approach is the *LR threshold* strategy, also known as the *KI threshold*, where we select only matches with an LR above a certain limit. This approach can potentially lead to a very large number of matches, depending on the threshold and the relationship we search for.

Profile centered

The *profile-centered* approach uses conditional distributions to find LR thresholds depending on the profile under considerations, hence the name. "Conditional" indicates that we compute the distribution of the LR given the profile of the candidate and the relationship we consider. We specify a value α, denoting the percentage of occurrences where the LR exceeds a certain threshold given the relationship is true. In other words, $(1-\alpha)$ corresponds to the quantile of LRs not exceeding this threshold.

The distributions may be found through conditional simulation, which given a large number of candidates may be computer intensive. We may also use the exact conditional distributions and find the necessary quantiles, though for a large number of markers the tables needed to store these distributions may be too large.

Conditional

Finally, the last approach employs Bayes's theorem to compute posterior probabilities for each candidate. This requires prior probabilities to be specified, which is not always straightforward. Once we have computed the complete posterior distribution, we select a number of candidates where the sum of their individual posteriors is equal to or greater than some threshold α. The name "conditional" refers to the use of Bayes's formula and the fact that each posterior is conditional on the LRs for all other candidates.

3.5 EXERCISES

The following exercises introduces how to use Familias (version 3.1.9) and above to search for relationships in different settings, e.g., identification in mass disasters, familial searching, blind searching etc. Software, input files and solutions (some in the form of videos) can be found following links from http://familias.name.

Exercise 3.1 DVI (a warm-up example). We consider a small DVI case with necessary data given in Figure 1.3. There are three victims, called V1, V2, and V3. Reference data have been obtained from two families, F1 and F2. There is one marker (M1) with equifrequent alleles 1, 2, 3, and 4. (We assume mutation rates equal to zero.) In the DVI module of `Familias`, the reference families are treated one-by-one sequentially. The hypotheses for family F1 are as follows:

H_1: V1 belongs to F1.
H_2: V2 belongs to F1.
H_3: V3 belongs to F1.
H_4: Some unknown person, V4, belongs to F1.

In other words, we consider four different possible hypotheses for the first family (F1). We start with some theoretical calculations before confirming the manual derivations in `Familias`.

(a) Find $L_i = \Pr(\text{data}|H_i),\ i = 1, 2, 3, 4$ by hand.
(b) Similarly, find $\text{LR}_i = L_i/L_4$.
(c) Why is it reasonable to scale the LRs versus H_4?
(d) We assume a flat prior—that is, $\Pr(H_i) = 0.25$. Calculate the posterior probabilities $\Pr(H_i \mid \text{data})$.
(e) The hypotheses for family F2 are as follows:

H_1: V1 belongs to F2.
H_2: V2 belongs to F2.
H_3: V3 belongs to F2.
H_4: Some unknown person, V4, belongs to F2.

Repeat (b) and (d) for the above hypotheses.
(f) Confirm the manual calculations in `Familias`. First, define the frequency database according to the specifications above.
(g) Open the DVI interface. *Hint*: `Tools->DVI module->Add Unidentified Persons`. Add persons and corresponding genotypes as specified for V1, V2, and V3 in Figure 1.3.
(h) Continue by specifying the reference families *Hint*: `Next` and then `Add`. The families are specified by first specifying the persons—that is, for F1 we define the typed father and mother, while for F2 we define the typed father. Add the genotypes as specified in Figure 1.3. Next add the pedigree specifying the necessary relationships between the typed persons and the missing person. *Hint*: `Add` in the `Pedigrees` section. Never mind the pedigree named `Reference pedigree` for now, we will return to the importance of this later.
(i) Perform a search with a `Threshold/Limit` of 0. *Hint*: `Next` and then `Search`.
(j) Confirm the manual calculations in (b), (d), and (e).

Exercise 3.2 Blind search (a warm-up example). The blind search interface is, as the name suggests, a tool to blindly search for predefined pairwise relationships in a data set. The interface currently allows the user to search for parent–child, siblings, half siblings, cousins, second cousins as well as direct matches. For the direct matching, `Familias` implements a special algorithm, which will be explored in more detail later. A blind search may be performed in connection with the DVI module—for example, to search a set of unidentified remains for direct matches or relationships within the data set. Another application may be to investigate a data set for unspecified relations before conducting a medical study or prior to creating a frequency database. The computations are swift and may be used to search large data sets for relations. We will first test the module on a smaller data set.

(a) Create one allele system with four alleles, 12, 13, 14, and 15, and allele frequencies uniformly distributed as 0.25.
(b) Define four males P_1, P_2, P_3, and P_4. Enter DNA data as $P_1 = 12/12$; $P_2 = 12/12$; $P_3 = 13/14$; $P_4 = 14/15$. *Hint*: You should define individuals in `Tools->Persons`.
(c) Enter the `Blind search` dialog. *Hint*: `Tools->Blind search`.
(d) Press `New search` and select `Direct-match` and `Siblings` as relationships. *Hint*: Hold down `Ctrl` to select multiple items in the list. Leave the `Match threshold` and θ at their default values. Set the remaining parameters (`Typing error`, `Dropout prob.`, and `Dropin`) to 0.
(e) Interpret the results.
(f) Confirm the LRs by manual calculations.
(g) Try different values for `Typing error`, `Dropout prob`, and `Dropin`. Specifically, change only one value to 0.1—that is, leave the other two at 0. Compare the results with what you expect (exact calculations are not expected here).

Exercise 3.3 DVI (an extended example). Consider the crash of a small plane with 10 passengers. We have obtained reference data from five different families. There are many steps and the exercise may take some time, but we encourage users to go through all steps as there is a lot to learn by doing this.

(a) In `Familias`, open the **Exercise3_3.fam** file, which contains frequency data for 23 autosomal markers.
(b) Enter the first step in the DVI module, `Add unidentified persons`. We may define individuals manually, similarly to the normal `Familias` procedure, though we prefer to import data from a file to skip as much manual input as possible. Import the file **Exercise3_3_pm.txt**. (`Familias` can import different files formats—for example, CODIS xml and tab-separated text files.)
(c) The file contains only eight unidentified remains. Discuss why this may be a realistic scenario, especially in larger-scale scenarios. How may this affect the calculations?

(d) Deselect `Use list` and enter 10 in the `Size` box. This is used to define the priors. We will not dwell on the discussion of priors now. Briefly we define the number of missing persons to be 10.

(e) Press `Next` to define reference families. We may now either define families manually or we may import them from a file. We will consider here two different alternatives. Define the first family manually by selecting `Add`. Enter a name for the family, *Family 1*.

(f) Import data for the persons in the family (a father). Import the file **Exercise3_3_am1.txt**. (*Note*: It is not necessary to first manually define the typed persons.)
If relevant, now is the time to define other persons included in the family, in the current family none. This may be untyped persons necessary to define the relations between the reference persons and the missing person(s). We will return to an example of this later.

(g) We continue with defining the relation between the defined person(s) and the missing person. (*Note*: Simply naming the person father/mother/brother, etc., does not define the relationships.) Select `Add` in the pedigree section to add a new pedigree. Name the pedigree appropriately, *Father*, and add the necessary relation between the reference person(s) and the missing person. Press `Close` and then `Close` again to return to the list of reference families.

(h) Define also a second family, where data is available for a brother of a missing person, by pressing `Add`. Enter a name, *Family 2*.

(i) Import reference persons from the file **Exercise3_3_am2.txt**

(j) Add necessary additional persons, untyped mother and father, and then define the reference person as the brother of the missing person. *Hint*: Add a pedigree as for Family 1 and specify that the brother and the missing persons share the same parents.

(k) Add the rest of the reference families by selecting the import option `Simple` and select the files **Exercise3_3_am3.txt, Exercise3_3_am4.txt**, and **Exercise3_3_am5.txt**. Change the names of the families to *Family 3*, *Family 4*, and *Family 5*. Also, check the persons and pedigrees in each imported family to make sure you know the relationships. Rename the pedigrees to reflect the defined relationships.

(l) Press `Next` and `Search` to start the matching. Select the threshold for a match to be reported. Enter 1.0, as we would rather obtain more matches at this stage and later remove matches which may be spurious.

(m) Interpret the results. Were all remains identified?

(n) Select a match and press `View match` to investigate the individual LRs for each system.

(o) We suspect there might be relatives among the unidentified persons. Enter the first step, `Add unidentified persons`, and select `Blind search`. Use your knowledge from Exercise 3.2 to perform a blind search for sibling relations. (Use 10 as the match threshold, leave all other options at the default.) How may the results be used in the DVI operation?

(p) Change the size of the accident in step (c) to 100 and see how this affects the priors in the current case. How does this, in turn, affect the posteriors? *Hint*: Perform a new matching to see the effect.
(q) *New information is added to the case. The first family, defined manually in (d), also contains a second missing person. The brother of the reference father is also missing. Try finding out how this could be solved with the means available in the DVI module.
(r) *Perform a new search; use the same match threshold as in (e).
(s) *Discuss the solution and other ways to improve the algorithm.
(t) Save the project.

Exercise 3.4 DVI (quick searching). Generally, pedigree structures may be complex, and thus calculations with a large number of unidentified individuals will be computer intensive. `Familias` implements functionality to speed up calculation in the DVI interface. This exercise will illustrate the features and how they are used.

(a) Open the file **Exercise3_4.fam**, which contains the final project from the previous exercise. Skip to the `Search` dialog in the DVI interface and press `Quick scan`. This will open a version of the `Blind search` dialog. Select `Parent-Child` and `Siblings` and match limit 10. Hit `Search`.
(b) Compare the current results with the results obtained in the previous exercise.
(c) What are the obvious benefits of our using the `Quick scan` function?
(d) In addition to the `Quick scan` function, `Familias` implements an algorithm to quickly compare reference families with postmortem samples using a zero mutation rate model. Enter the advanced settings, `File->Advanced`, and select/deselect `Quick search` to activate/deactivate this feature.
(e) Make sure the `Quick search` is selected and enter 1 as the number of allowed mismatches. Return to the `Search` dialog in the DVI interface and perform a search with a match threshold of 0.0001.
(f) Save the results with use of the `Export list` function.
(g) Return to the advanced settings and change the number of allow mismatches to 2.
(h) Perform a new search and compare the results with the ones obtained in (f).
(i) In light of the results, discuss "good" values for the number of mismatches and the benefits of using the quick search features.

Exercise 3.5 DVI (a simulated example). This exercise is divided into two parts:

1. Simulate data in `Familias` and prepare it for input to the DVI module. (Requires `Excel` or similar software).
2. Use the DVI module on the simulated data.

The exercise deals with a larger data set of simulated pairs of sisters with real frequency data for 23 autosomal STR markers. As described, the first part of the exercise deals with how to simulate relationships in Familias and how to use the genotype data from the simulations. Skip to (h) for DVI exercises only. Parts (b) through (g) are particularly useful for readers interested in simulating data for validation purposes and to understand some of the import formatting recognized by Familias.

(a) Open the database from the file **Exercise3_5.fam**, containing frequency data for 23 autosomal STR markers. Briefly explore the database before continuing to make sure you are familiar with the markers and parameters.

(b) Enter the Advanced dialog (*Hint*: File->Advanced) and make sure Save genotype data is selected and Save complete data is deselected. Press Save.

(c) Define the persons necessary to simulate a pair of full siblings—that is, Mother, Father, Sister1, and [Sister]. It is important that you name the second sibling as indicated—that is, [Sister]. In later steps Familias will recognize this as a relationship indicator and construct pedigrees automatically.

(d) In the Pedigree dialog, create two pedigrees, the first specifying Sister1 and [Sister] as full siblings, and the second one without any relations.

(e) Press Simulate. Specify that we will have genotype data for Sister1 and [Sister]. Enter 100 simulations and set the seed to 12345. Also, select to save the raw data.

(f) Press Simulate and save the genotype data to a text file. (Leave Familias open.)

(g) The following steps prepare the output for import into the DVI module of Familias:

1. Open the simulated genotype data in Excel, or similar software. (If the alleles appear as dates, consult http://familias.no/english/help/.)
2. Save the file as a new file in the xls or xlsx format for better compatibility.
3. Select the first row and select Filter. *Hint*: It is usually found in the Data tab.
4. In the first column (True ped) select to only view Ped 1 (or the name you gave to the full siblings pedigree) and in the second column select to view only Sister1.
5. Now, copy all the filtered data into a new file. *Hint*: Ctrl+a followed by Ctrl+c followed by Ctrl+n followed by Ctrl+v. Remove the first column—that is, information about the pedigree.
6. In the first column, select the first occurrence of Sister1. Fill down (https://exceljet.net/keyboard-shortcuts/fill-down-from-cell-above) such that the numbering is Sister1, Sister2,. . . ,Sister100.
7. Save the new formatted file as **Exercise3_5_pm.txt** and make sure the format is a tab-separated text file. (This file now contains data for 100 unrelated individuals serving as the postmortem data later.)
8. Return to the original Excel document and remove the filtering. In the first column select again to view only Ped 1, and in the second column select to view only [Sister]. With the same procedure as in 5, copy the filtered data to a new document. In the first column

in the second row, rename `Ped 1` as `Family1`. Fill down such that the numbering becomes sequential—that is, Family1, Family2,. . . ,Family100.

9. Save the data as **Exercise3_5_am.txt**. This file now contains data for 100 reference sisters serving as the antemortem data later

(h) Open the DVI interface and import the file **Exercise3_5_pm.txt** containing the data for 100 unrelated individuals serving as the set of unidentified remains. *Hint*: If you did not perform the simulation steps, see (a) before continuing.

(i) Run a blind search and search for siblings with a match threshold of 10. (Leave all other parameters at their default values.)

(j) Comment on the results.

(k) Continue to specify reference families. We will use the `Data only` import option, allowing a simple import. Select and import the file **Exercise3_5_am.txt**. This will import 100 reference families with a sister as the reference person. Explore the families and see how the import has automatically detected the relationships.

(l) Perform a search with a match threshold of 10. Comment on false positives/negatives and investigate spurious matches by use of the `View match` button. *Note*: SisterX goes with Family X, where X is an integer from 1 to 100.

Exercise 3.6 Familial searching (warm-up example). This exercise introduces the *Familial searching* module in a simple case. First some calculations are done by hand which are later checked with `Familias`. Consider a marker L1 with alleles 12, 13, 14, and 15, all with frequency 0.25, and a database of convicted offenders consisting of four individuals with genotypes $P_1 = 12/12, P_2 = 12/13, P_3 = 13/14$, and $P_4 = 14/15$. For simplicity, we disregard complicating factors such as mutation, theta correction, drop-in, dropout, and typing error.

(a) There is a stain $S_1 = 12/13$, assumed to be from one individual. Consider:

H_i: $S_1 = P_i$,
H_0: S_1 is unrelated to P_i,

and calculate $\text{LR}_i = \Pr(\text{data} \mid H_i)/\Pr(\text{data} \mid H_0)$. (Answer: $\text{LR}_2 = 1/(2 \times 0.25 \times 0.25) = 8$ and $\text{LR}_i = 0$ for $i \neq 2$.)

(b) Repeat the above calculations for $H_i : S_1$ is a child of P_i. (Answer: $\text{LR}_1 = 2, \text{LR}_2 = 2, \text{LR}_3 = 1, \text{LR}_4 = 0$.)

(c) There is a stain $S_2 = 12/13/14/15$ assumed to be from two contributors. Consider the following:

H_i: The stain comes from P_i and an unrelated individual not in the database.
H_0: The stain comes from two unrelated individuals not in the database.

Calculate $\text{LR}_i = \Pr(\text{data} \mid H_i)/\Pr(\text{data} \mid H_0)$. Answer: We first find the likelihoods:

$$\Pr(\text{data} \mid H_0) = 24 \times p_{12}p_{13}p_{14}p_{15} = \frac{3}{32},$$

$$\Pr(\text{data} \mid H_1) = 0,$$

$$\Pr(\text{data} \mid H_i) = \frac{1}{8}, \quad i = 2, 3, 4.$$

Therefore, $\text{LR}_1 = 0$, $\text{LR}_2 = \text{LR}_3 = \text{LR}_4 = \frac{4}{3}$.

(d) Consider the following:

H_i: The stain comes from the son of P_1 and an unrelated individual (NN) not in the database.

H_0: The stain comes from two unrelated individuals not in the database.

Calculate $\text{LR}_1 = \Pr(\text{data} \mid H_1)/\Pr(\text{data} \mid H_0)$. Answer: The son must be 12/13, 12/14, or 12/15, and so

$$\begin{aligned}\Pr(\text{data} \mid H_1) &= \Pr(\text{son} = 12/13, \text{NN} = 14/15 \mid \text{father} = 12/12)\\ &\quad + \Pr(\text{son} = 12/14, \text{NN} = 13/15 \mid \text{father} = 12/12)\\ &\quad + \Pr(\text{son} = 12/15, \text{NN} = 13/14 \mid \text{father} = 12/12)\\ &= 6p_{13}p_{14}p_{15}.\end{aligned}$$

Therefore, $\text{LR}_1 = \frac{1}{4p_{12}} = 1$.

(e) Consider the above problem with $P_2 = 12/13$ replacing P_1 and calculate $\text{LR}_2 = \Pr(\text{data} \mid H_2)/\Pr(\text{data} \mid H_0)$. Answer: The son must be 12/13, 12/14, 12/15, 13/14, or 13/15, and so

$$\begin{aligned}\Pr(\text{data} \mid H_2) &= \frac{1}{2}(p_{12} + p_{13})2p_{14}p_{15}\\ &\quad + 2p_{13}p_{14}p_{15} + 2p_{12}p_{14}p_{15}\\ &= 3p_{12}p_{14}p_{15} + 3p_{13}p_{14}p_{15}.\end{aligned}$$

Therefore, $\text{LR}_2 = \frac{p_{12}+p_{13}}{8p_{12}p_{13}} = 1$.

(f) Next, the above calculations are confirmed with `Familias`.

1. Start `Familias` and define the marker L1.
2. Enter `Tools->Familial searching`: Enter name P_1 and `Add`. Add genotype data by selecting system L1. The manual input deviates from other parts of the program to allow for mixtures. Select `Allele 1` and 12. Select `Allele 2` and 12. Press `Add` to add the observations.
3. Define P_2, P_3, and P_4 similarly.
4. Enter `Next` and define the stain S_1.
5. Enter 0 for all search options—that is, `LR threshold` and the other parameters should be set to 0.
6. Select `Parent-child` and `Direct match` in the upper right corner.
7. Enter `Next` and press `Search`. Confirm the answers in (a) and (b).

8. Return to the previous dialog, where we defined the stain. Import a mixture from the file **Exercise3_6.xml**. The sample is the previous mixture of two contributors: 12/13/14/15.
9. Confirm the previous calculations.

Exercise 3.7 Looking for the relative of a stain. The following exercise will explore the *Familial searching* module more thoroughly.

(a) Start by importing the frequency database from the file **Exercise3_7.fam** and explore the contents.
(b) The database containing convicted offenders and traces from previous crimes is contained in the file **Exercise3_7db.txt**. Import the file into the familial searching interface. *Hint*: `Tools->Familial searching`. In total there should be 1000 elements, genotyped for different sets of markers. The latter may be common as different marker kits may have been used throughout the history of the database.
(c) Perform a blind search on the database; use the option *Direct match*. Use default parameter values. Be patient, the search may take a while. How many comparisons are performed?
(d) Comment on the results and whether the values of the dropout, drop-in, and typing error parameters (default values) are appropriate.
(e) Can a blind search be performed on a database of say 5,000,000 elements? Compute the number of comparisons necessary for such an operation.
(f) Press `Next` and continue with importing a batch of traces from some crime scenes. Import data from the file **Exercise3_7tr.txt**.
(g) Specify that you wish to search for `Direct match` first. Let *Dropin* = 0.01, *Dropout* = 0.05, and *Typing error* = 0.001. Set the match threshold to 1.
(h) Perform a search and comment on the results.
(i) Return to the previous dialog to specify a new search. Now select the *Parent–Child* and *Sibling* relations. Set the match threshold to 10 and leave the other settings at their default values.
(j) Perform a new search and sort the matches by means of the `Sort` button. Select a subset of the matches to explore further. Use the `Subset` button and select the *top-k* method. Enter 10 as the number of matches to select. Review the results.
(k) Press `Search` to redo the search. Select the *LR threshold* method in the subset feature. Enter 100 and apply. Review the results.
(l) What are the benefits and downsides of the use of the *top-k* method compared with the *LR threshold* method?
(m) `Familias` furthermore implements another subsetting method, *profile centered*. The algorithm will perform conditional simulations based on the candidate (matching) profile and find an LR threshold based on the results. Perform the search again and apply the *profile centered* method with the `alpha` parameter set to 0.9. Review the results and compare them with the ones obtained with the other two subset methods.

(n) *Try explaining the benefits of the use of the *profile centered* method and the meaning of the `alpha` parameter. What will happen for low values of `alpha`? Confirm your answer in `Familias`.

(o) Now, return to the search options and specify that you wish to search for direct match and parent–child relationships and scale against `Cousins`. In addition, specify the value of F_{st} to be 0.02. When may these specifications be relevant? (Leave the other parameters at their default values.)

(p) Perform a search with the new settings and comment on the results. Compare them with the results obtained in (j).

Exercise 3.8 *Mixtures and relatives. This exercise will demonstrate the *familial searching* interface when the evidence is a mixture. The ideas were briefly touched on in Exercise 3.6, and we will consider here some more theoretical points as well as confirm these, when possible, in `Familias`.

Beginning with the basics, for mixtures we may consider hypotheses:

H_1: The stain S_1 is a mixture of $N-1$ unrelated individuals and the unrelated profile of interest, P_1.

H_2: The stain S_1 is a mixture of N unrelated individuals. P_1 is not in the stain and is unrelated to all N individuals.

In the current exercise we consider $N = 2$.

(a) We define one allele system with alleles and frequencies as indicated in Table 3.4 (mutation rates are set to zero).

(b) *The stain S_1 is genotyped as 12/13/14, while the genotype of P_1 is typed as 12/13. Compute by hand the LR comparing H_1 with H_2.

(c) Prepare the data in `Familias` and compute the LR for the *Direct-match* between S_1 and P_1 in the `Familial searching` module of `Familias`. (Set all parameters to zero.)

We can extend the above-mentioned hypotheses to include a number of alternative hypotheses about relatedness, which are relevant in the context of *familial searching*. We may specify the following:

H_3: The stain S_1 is a mixture of $N-1$ unrelated individuals and the relative R_1 of P_1.

Table 3.4 Allele Frequencies for a Genetic Marker

Alleles	Frequencies
12	0.1
13	0.2
14	0.3
15	0.2
16	0.1
17	0.1

(d) *Use the same specifications as in (a), and compute by hand the LR comparing H_3 and H_2 (as specified above) and under the assumption that the child of P_1 may be in S_1.

(e) *Confirm the calculations in `Familias`.

(f) **Compute by hand the LR comparing H_3 and H_2, but assume the untyped mother of child is in S_1. In other words:

H_2**:** The stain S_1 is a mixture of a mother and her child. P_1 is unrelated to the child.

H_3**:** The stain S_1 is a mixture of a mother and her child. P_1 is the father of the child.

Note: The LR cannot be confirmed in `Familias`.

(g) *Consider instead that the relative, R_1, is the brother of P_1. Compute by hand the LR comparing H_3 and H_2 where the former hypothesis specifies that P_1 and R_1 are brothers.

(h) *Confirm the calculations in `Familias`.

(i) *Add three more stains (S_2, S_3, S_4) with genotypes 12/13, 12/12, and 12/14, respectively. Given that these are stains with a single contributor and that we are still searching for siblings of P_1 in the stains, try predicting what the LR will be for these different genotype constellations.

(j) *Confirm your ideas in `Familias` and compare the results with the ones you obtained in (h).

Exercise 3.9 Further use of the blind search function. In this exercise we will take a closer look at how to use the blind search function in a different setting. We will consider the scenario where we wish to create a new database for a specific population.

(a) In `Familias`, open the `Create database` function.
Hint: `File->Create database`.

(b) Import the file **Exercise3_9.txt** containing output for 210, supposedly unrelated individuals, sampled from a small subpopulation. *Hint*: Use the `Import` button. The data is prepared in the standard `Familias` format (i.e., tab-separated text file), though direct output from, for example, `Genemapper` is accepted as well.

(c) The first step before creating the database is to ensure that the individuals are truly unrelated. Open the `Blind search` function from within the `Create database` dialog with the `Check data` button.
The blind search will create a temporary database, based on the 210 individuals. The statistical calculations may be biased, but will provide a general idea of any relationships in the data set.

(d) Perform a search for direct matches and parent–child relations. Use a match limit of 1000 and set the F_{st} correction to 0.05 (Leave all other parameters at the default values).

(e) Comment on the results. Export the list by means of the `Export list` function.

(f) Return to the previous dialog and remove all samples with an LR above 100,000. (*Note*: It is sufficient to remove one of the samples in a match.)
In a real situation it may be that we desire a more specific investigation into why the samples match, but for now remove the sample with the lowest number.
(g) Create a summary of some statistical parameters. *Hint*: Use the `Statistics` button to export relevant information to a file. Review the file in spreadsheet software such as `Excel`. (For problems with `Excel`, see http://familias.name.)
(h) Create the database and save the file. Explore the database created.

Exercise 3.10 *Direct matching. In this exercise we take a closer look at the direct searching algorithm. We will consider data for one autosomal marker, vWA. We wish to compare the following hypotheses:

H_1: Two profiles, G_1 and G_2, belong to the same individual.
H_2: G_1 and G_2 belong to two unrelated individuals.

LR is computed according to

$$\begin{aligned} \text{LR} &= \frac{\Pr(G_1, G_2 \mid H_1)}{\Pr(G_1, G_2 \mid H_2)} \\ &= \frac{\sum_{i=1}^{N} \Pr(G_{\text{true},i})\Pr(G_1 \mid G_{\text{true},i})\Pr(G_2 \mid G_{\text{true},i})}{\Pr(G_1)\Pr(G_2)}, \end{aligned} \tag{3.7}$$

see also Section 3.3.2 for details.

The direct matching feature allows any two profiles to be matched against each other, and we may calculate the probability that they originate from the same latent profile ($G_{\text{true},i}$), where the latent profile is a priori unknown and we have to sum over all possibilities. We compute transition probabilities, $\Pr(G_1 \mid G_{\text{true},i})$ and $\Pr(G_2 \mid G_{\text{true},i})$. For the current exercise we can use the probabilities listed in Table 3.5, where d is the dropout probability, c is the drop-in parameter, and e is the typing error

Table 3.5 Transition Probabilities Between Genotypes, $j = 1, 2$

G_j	$G_{\text{true},i}$	$\Pr(G_j \mid G_{\text{true},i})$
a/a	a/a	$(1-d^2)(1-e)(1-c)$
a/a	a/b	$e+(1-d)d(1-c)(1-e)$
a/a	b/b	$e(1-d)^2(1-c)$
a/a	b/c	$e(1-d)^2(1-c)$
a/b	a/b	$(1-e)(1-c)(1-d)^2$
a/b	a/a	$e+(1-e)(1-d)^2cp_b$
a/b	a/c	$e(1-c)(1-d)^2+d(1-d)(1-e)cp_b$
a/b	c/c	$e(1-c)(1-d^2)$
a/b	c/d	$e(1-c)(1-d)^2$

probability. (*Note*: The table displays a simplified version of probabilities, neglecting events such as two dropouts and one drop-in occurring at the same time.)

(a) Open the file **Exercise3_10.fam** containing a frequency database for the marker vWA.

(b) Import data for 10 individuals from the file **Exercise3_10.txt**. *Hint*: Use `Tools->Case-related DNA data`. The file contains no information on gender, and `Familias` will display a warning, which may be ignored.

(c) In the `Blind search` dialog, press `New search`. Select *Direct-match* and specify `Typing error`, `Dropin parameter`, and `Dropout prob.` to be zero. Use a match threshold of 1.

(d) View the match between P_9 and P_{10}. Compute the LR by hand.

(e) Perform a new search, now setting `Dropout prob.` to 0.1. (Leave the other parameters as previously stated.)

(f) Explore the results; press `View match` to investigate the profiles and matches more closely.

(g) *Calculate the match between the individuals named P_1 and P_2 using Equation 3.7.

(h) *In words, explain why the LR becomes so high even though the profiles are different. Explain also why the match between P_9 and P_{10} gets a lower LR than in (d).

(i) Repeat (e) but change the parameters to `Typing error`=0, `Dropin prob.`=0.1, and `Dropout prob.`=0.

(j) *Calculate the match probability for the individuals named P_1 and P_2 by hand.

(k) We may actually compare H_1 with some other hypothesis about relationships—for example,

H_2: G_1 and G_2 belong to two brothers.

This may, for example, be relevant in a criminal investigation, where the defendant states that *It was my brother who did it!*
Perform a new search, with the same parameters as in (e), but specify that we wish to scale versus siblings.

(l) Comment on the new results.

(m) *Use the results in (g), and calculate the LR by hand for the match between P_1 and P_2. (*Note*: computations under H_2 do not consider dropout, drop-in, and typing errors.)

CHAPTER

4 Dependent markers

CHAPTER OUTLINE

There are important and practical problems that cannot be solved with the number of independent short tandem repeat (STR) loci available and conventionally used. For instance, it is normally not possible to distinguish half siblings reliably from full siblings [80] when parental DNA profiles are unavailable. There are also symmetric problems that cannot be resolved for any number of unlinked autosomal markers: the likelihood ratio (LR) comparing a half-sibling relation with an avuncular relation will always be 1 as pointed out in the "Allowing linked markers" section of Chapter 6. This requires more data, typically linked, dependent, markers.

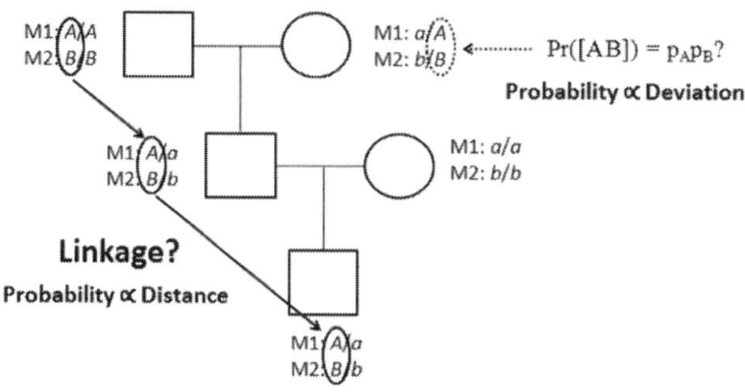

FIGURE 4.1

The concepts of linkage and linkage disequilibrium (LD). Linkage is depicted as the haplotype [A B] being transmitted as a unit in the pedigree, while LD relates to the probability of the haplotype [A B] in the population.

Dependency may be a result of closely located loci, both within pedigrees (linkage) and in the population (allelic association or linkage disequilibrium, LD); see Figure 4.1. This requires new methods and implementations compared with those discussed so far for likelihood calculations. This chapter will discuss some important concepts connected to dependency problems with genetic markers. The programs `FamLink` [5] and `FamLinkX` [6] will be used to illustrate efficient implementations to perform likelihood calculation; the latter is specifically designed to deal with X-chromosomal markers.

Some parts of this chapter contain more statistical theory, and the essential parts are explained such that readers with little mathematical background can follow the discussion, and again most of the exercises do not require all the theory.

4.1 LINKAGE

4.1.1 RECOMBINATION

Recombination is a biological phenomenon occurring during meiosis, that is the cell division of sex cells. Recombination, or crossover, occurs when homologous chromosomes exchange segments and the resulting new chromosomes are a random combination of the original two chromosomes. The process vastly increases the biological diversity; see Figure 4.2 for a simplified illustration. It is governed by elaborate mechanisms of the cell, and this book does not provide all the details; the interested reader should instead consult literature in basic genetics, say [81]. Moving along a chromosome, we may address the probability of observing a crossover between any two given spots—that is, the recombination probability. Intuitively, the further away two spots are located, the higher this probability, or fraction, is.

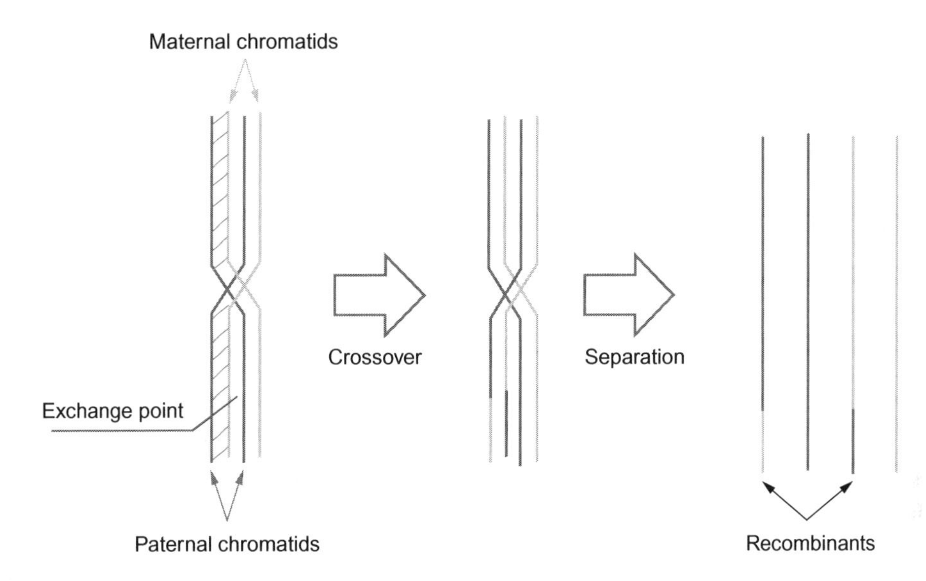

FIGURE 4.2

Chromosomal crossover. The pairing of the duplicated chromosomes during meiosis and the exchange of a part of segments for two of the chromatids are illustrated. The final result is two recombinant chromosomes and two non-recombinants.

Isolating two loci of interest, we are looking for the probability that an odd number of crossovers occurs between them as an even number cannot be observed unless we consider information from other loci. In other words, a recombination has occurred between two loci if an odd number of crossover events is recorded. Without specific consideration of the biological process, Haldane[1] proposed a Poisson process to illustrate the procedure of moving along the chromosome. The number of crossovers X is assumed to be approximately distributed as Po(d) with density function

$$\Pr(X = k) = \frac{\mathrm{e}^{-d} d^k}{k!}.$$

We need to sum over all odd numbers of k, and we may write

$$\sum_{m=0}^{\infty} \frac{\mathrm{e}^{-d} d^{2m+1}}{(2m+1)!}, \tag{4.1}$$

using $k = 2m + 1$ to obtain all odd numbers.

Furthermore, using Equation 4.1, we can derive the Haldane mapping function—that is, the probability that we observe an odd number of crossovers between any two loci:

[1] http://en.wikipedia.org/wiki/J._B._S._Haldane

$$\rho = \sum_{m=0}^{\infty} \frac{e^{-d} d^{2m+1}}{(2m+1)!} = e^{-d} d + e^{-d} \frac{d^3}{3!} + \cdots = e^{-d} \left(d + \frac{d^3}{3!} + \cdots \right) = \frac{1 - e^{-2d}}{2}, \quad (4.2)$$

which relates the recombination rate,[2] denoted here by ρ, to the genetic distance (d) in morgans (M). (The last step is derived through Maclaurin[3] expansions). From Haldane's mapping function, we see that the recombination rate can never exceed 0.5 as the function has its maxima when d approaches infinity and

$$\lim_{d \to \infty} \frac{1 - e^{-2d}}{2} = 1/2.$$

It is known that a crossover creates interference—that is, given that a crossover has occurred, it is less probable that another will occur in the immediate vicinity. A variety of other mapping functions, accounting for interference, exist [81], with one of the most commonly used being Kosambi's function:

$$\rho = \frac{e^{4d} - 1}{2(e^{4d} + 1)}.$$

See Ott [82] for a more comprehensive discussion on mapping functions.

Recombination rates may be calculated with use of maps of genetic distances—for example, with use of the above-mentioned mapping functions and an online map of genetic positions.[4] To obtain more accurate estimates we use large multigenerational families where individual haplotypes can be traced and recombinations can be observed. We may then use, for instance, maximum likelihood algorithms to compute an estimate of the rate.

4.1.2 INTRODUCTION TO CALCULATIONS

We will see that accounting for recombination between linked markers is not always necessary. Some cases are influenced, while others are not. A general rule of thumb is given by Gill et al. [83] and states that two or more meioses must separate two individuals in order for linkage to potentially affect the calculations. In addition, for some scenarios where the above rule is obeyed, linkage will not affect the calculations as certain genotype constellations cancel out the recombination rate. In addition, if association is present between alleles at the markers under consideration (see Section 4.2), linkage may have an affect even in regular paternity cases.

Theoretical derivations of the algebraic formulas may be approached in different ways with use of different notation. We present some ideas but there are others, not covered here.

[2]Different notation for the recombination rate is often seen in the literature—for example, r or θ, while we use ρ in order not to create confusion with other parameters used throughout the book.
[3]The Maclaurin expansion is a special case of the Taylor expansion, which is used in mathematics to represent a function as a sum of terms.
[4]Rutgers map: http://compgen.rutgers.edu/mapinterpolator

Example 4.1 Simple example: paternity case with linkage. To better illustrate linkage, we start with a simple paternity case (duo), with hypotheses as discussed in Section 2.3.1. Furthermore, consider two loci L_1 and L_2, closely located on the same chromosome. Assume the alleged father is heterozygous 12/13 and 21/23 for the two markers and the child is heterozygous 12/14 and 21/22. Disregarding complications such as mutations, silent alleles, and dropouts, we have

$$\begin{aligned}\text{LR} &= \frac{\Pr(\text{data} \mid H_1 = \text{paternity})}{\Pr(\text{data} \mid H_2 = \text{nonpaternity})} \\ &= \frac{H_{14,22}\left(0.5 \times 2H_{12,21}H_{13,23}(1-\rho) + 0.5 \times 2H_{12,23}H_{13,21}\rho\right)}{\left(2H_{12,21}H_{14,22} + 2H_{12,22}H_{14,21}\right)\left(2H_{12,21}H_{13,23} + 2H_{12,23}H_{13,21}\right)},\end{aligned} \tag{4.3}$$

where $H_{x,y}$ denotes the haplotype frequency for the combination of allele x at marker L_1 and allele y at marker L_2, and ρ denotes the recombination rate. As illustrated in Figure 4.3, given H_1, the haplotype that has been transmitted from the alleged father must be [12 21]. In turn, this imposes restrictions on the haplotypes for the child, as she must have obtained [12 21] from the father and [14 22] from the mother. However, this does not necessarily mean that [12 21] is the original haplotype for the father. In the formula, we consider the loci as a unit; in other words, we must consider all possible haplotypes for untyped individuals (and also typed individuals unless the data indicate phase[5]). This is obvious from the denominator in the above formulation, where we need to sum over the two possible haplotype setups for both individuals. In later examples, we will see other approaches where summation over ambiguous haplotypes is not always necessary.

For now we assume there is no association between the alleles (see Section 4.2), and we have $H_{x,y} = p_x \times p_y$. The formula above simplifies to LR $= 1/(16p_{12}p_{21})$; see also Exercise 4.1. In other words, the recombination rate will disappear and the LR depends only on the allele frequencies.

As discussed in Example 4.1, summation over haplotypes is not always a good approach for likelihood calculations. Consider, for instance, the case where we

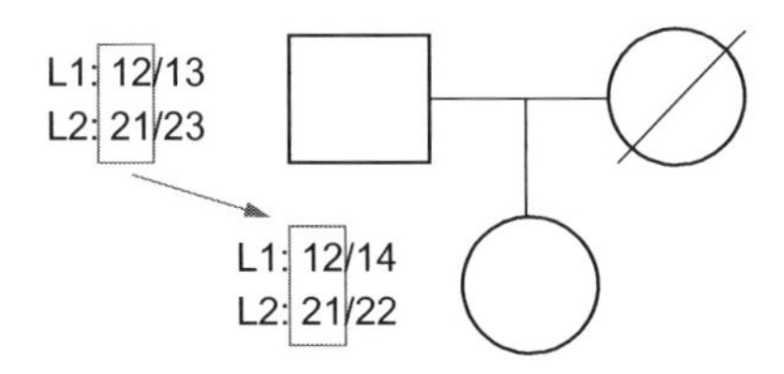

FIGURE 4.3

A duo case with a pair of linked markers.

[5]Knowledge of phase is equivalent to knowing which part of a genotype belongs to the maternal or paternal chromosome. In some special circumstances, the phase can be known, as when all loci under consideration are homozygous or all loci but one are homozygous

have N consecutive loci that we need to model as a unit; then, the number of potentially different haplotypes we need to investigate grows with 2^N. So even in a simple paternity case the computations for 20 linked markers would potentially require $2^{20} \times 2 \approx 2{,}000{,}000$ different haplotypes to be considered. There are some special circumstances when the haplotype approach may be used. For instance, if the markers are assumed to be completely linked—that is, $\rho = 0$ — then the number of haplotypes will be reduced. Consider again the case in Example 4.1 and that $\rho = 0$. Given H_1, the numerator now reduces to $H_{14,22}H_{12,21}H_{13,23}$, as the alleged father is obliged to have the haplotype [12 21]. Even so, many situations still require extensive summation for untyped individuals.

In the next example, we consider an alternative approach that uses "conditioning" on identical-by-descent (IBD) states, i.e., computations use the conditional probabilities given these states. This will later generalize to the Lander–Green algorithm [84] in Section 4.1.3.

Example 4.2 A sibling case with linked markers. We next visit a more complicated example dealing with linked markers. Let us consider the following hypotheses:

H_1: Two persons, P1 and P2, are related as full siblings.
H_2: P1 and P2 are unrelated.

Furthermore, consider two linked markers, L_1 and L_2. Genotype data are given in Table 4.1.

Use of the haplotype approach, as introduced in Example 4.1, would require extensive summations for the different possible haplotypes the parents of the siblings could have. Instead, we can derive the theoretical formulas for the likelihoods with extensions of the notation in Equation 2.6. First considering the evidence under H_1, we get (disregarding mutations and other complications)

$$\begin{aligned}
L(H_1) &= \Pr(\text{data} \mid H_1) \\
&= \Pr(I_{L_1} = 0)\Pr(\text{data} \mid I_{L_1} = 0)\left(\Pr(I_{L_2} = 0 \mid I_{L_1} = 0)\Pr(\text{data} \mid I_{L_2} = 0)\right) \\
&\quad + \Pr(I_{L_1} = 1)\Pr(\text{data} \mid I_{L_1} = 1)\left(\Pr(I_{L_2} = 0 \mid I_{L_1} = 1)\Pr(\text{data} \mid I_{L_2} = 0)\right) \\
&= 0.25 \times 4p_{19}p_{21}^2p_{25}\left(\rho^4 + (1-\rho)^4 + 2\rho^2(1-\rho)^2\right)p_{12}^2p_{14}^2 \\
&\quad + 0.5 \times p_{19}p_{21}p_{25}\left(2\rho(1-\rho)^3 + 2\rho^3(1-\rho)\right)p_{12}^2p_{14}^2,
\end{aligned}$$

where $\Pr(I_{L_1} = x)$ denotes the prior probability of two siblings sharing x alleles IBD at L_1. $\Pr(I_{L_2} = x \mid I_{L_1} = y)$ denotes the probability that y alleles are IBD at L_2 given

Table 4.1 Genotype Observation for a Pair of Linked Markers

	P1	P2
L_1	19/21	21/25
L_2	12/12	14/14

that x alleles are IBD at L_1. For instance, $\Pr(I_{L_2} = 0 \mid I_{L_1} = 0) = \rho^4 + (1-\rho)^4 + 2\rho^2(1-\rho)^2$ is found as the events when an even number of meioses occur—that is, when zero, two, or four recombinations occur, the first and last can occur in only one way, while two recombinations can occur in two different ways. Given that $\rho = 0.5$—that is, the markers are completely unlinked—we see that $\Pr(I_{L_2} = 0 \mid I_{L_1} = 0) = 0.5^4 + 0.5^4 + 2 \times 0.5^4 = 0.25$. We further note that the siblings can share at most only one allele IBD at L_1, and thus $\Pr(\text{data} \mid I_{L_1} = 2) = 0$. Given the data at L_2, the siblings must share no alleles IBD, and thus both $\Pr(\text{data} \mid I_{L_2} = 2) = 0$ and $\Pr(\text{data} \mid I_{L_2} = 1) = 0$. The likelihood given H_2 is found as

$$\begin{aligned} L(H_2) &= \Pr(\text{data} \mid H_2) \\ &= \Pr(I_{L_1} = 0)\Pr(\text{data} \mid I_{L_1} = 0)\Pr(I_{L_2} = 0 \mid I_{L_1} = 0)\Pr(\text{data} \mid I_{L_2} = 0) \\ &= 1 \times 4p_{19}p_{21}^2p_{25} \times 1 \times p_{12}^2p_{14}^2, \end{aligned}$$

which is only a product of allele frequencies as H_2 involves no meioses. Finally, computing the LR, we get

$$\begin{aligned} \text{LR} &= \frac{\Pr(\text{data} \mid H_1)}{\Pr(\text{data} \mid H_2)} \\ &= \cdots = \frac{p_{21}\left(\rho^4 + (1-\rho)^4 + 2\rho^2(1-\rho)^2\right) + 0.5\left(2\rho(1-\rho)^3 + 2\rho^3(1-\rho)\right)}{4p_{21}} \\ &= \frac{\rho^4 + (1-\rho)^4 + 2\rho^2(1-\rho)^2}{4} + \frac{0.5\left(2\rho(1-\rho)^3 + 2\rho^3(1-\rho)\right)}{4p_{21}}. \end{aligned}$$

We may plot the LR as a function of ρ for some values of p_{21}; see Figure 4.4. As shown in the figure, the smaller the value of p_{21}, the larger the impact of linkage.

These two first examples illustrate different approaches for likelihood calculations that use linked markers, the first using haplotype notation, the second using notation similar to that presented for the more general formulation, which is described in Section 4.1.3. While the first approach may seem intuitive, the latter approach and the general formulation are recommended and will be used throughout the rest of the book.

4.1.3 GENERALIZATION AND THE LANDER–GREEN ALGORITHM

In 1987, Lander and Green [84] introduced an algorithm to effectively compute likelihoods based on pedigree structures and genetic marker data. The algorithm embodies the ideas of hidden Markov chains to handle marker dependency (linkage); see Figure 4.5 for an illustration.

The property sought is the fact that given the value of a node in the Markov chain, subsequent nodes are independent of previous nodes. In other words, we have to consider only linkage between adjacent loci, and thus the computations do not grow

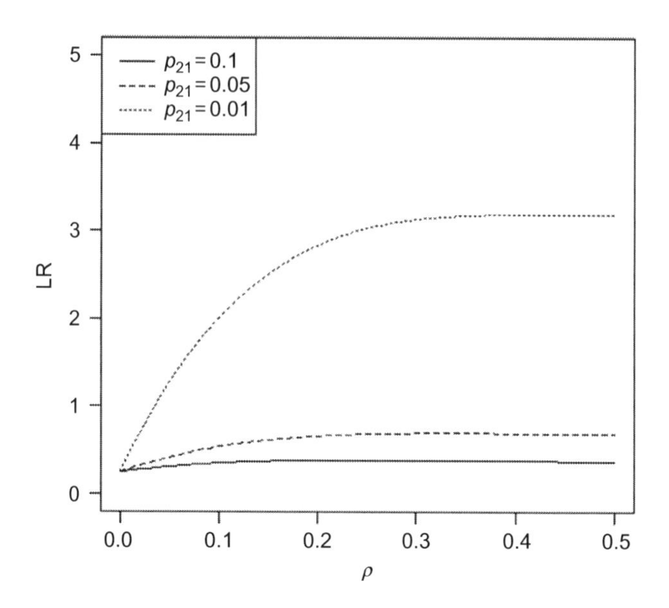

FIGURE 4.4

The LR as a function of ρ for a sibling case.

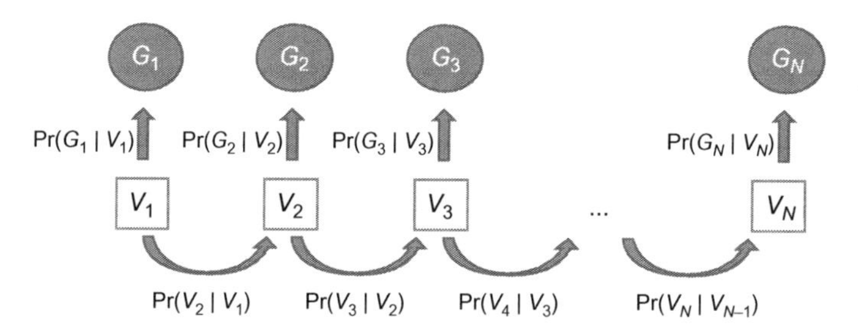

FIGURE 4.5

A hidden Markov chain for genetic markers. The squares denote the hidden inheritance patterns ($V_1,\ldots,V_N$) and the circles denote the observed genotypes ($G_1,\ldots,G_N$).

considerably in complexity with the number of markers. A version of the Lander–Green algorithm is provided below:

$$L = \sum_{V_1} \cdots \sum_{V_N} \Pr(V_1) \prod_{i=2}^{N} \Pr(V_i \mid V_{i-1}) \prod_{i=1}^{N} \Pr(G_i \mid V_i), \tag{4.4}$$

which in fact is a shorter version of the complete Lander–Green algorithm where the probabilities relating phenotypes such as disease status to inheritance patterns

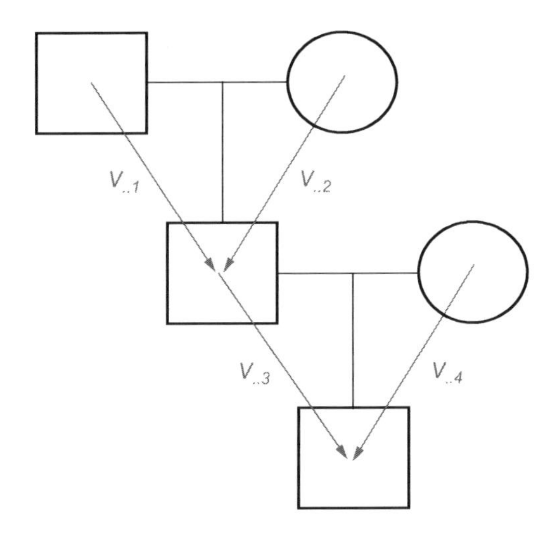

FIGURE 4.6

All meioses in a pedigree, denoted $v_{..1}, v_{..2}, v_{..3}, v_{..4}$.

is omitted. We may first dissect the formula using simple wording and start by specifying a list of all meioses in a given pedigree; see Figure 4.6.

Meiosis is defined here as the division of sex cells—that is, it is described by transmissions in a pedigree. Looking at a paternity case (and considering autosomal markers), with data for both the alleged father and the mother, we must have two meioses, one on the paternal side and one on the maternal side. For a case of two full siblings, we have to account for in total four meioses, the first two specifying whether the grandpaternal or grandmaternal allele has been transmitted from the father or the mother, respectively, for the first person, while the last two specify the same for the second person. In practice, we denote this with binary indicators: 0 if the grandmaternal allele has been inherited and 1 if the grandpaternal allele has been inherited. The binary vector specifying one inheritance pattern would look like [0 0 0 0], indicating two alleles IBD.

In the Lander–Green algorithm, V_i defines the complete list of all possible meioses—that is, the inheritance space—at marker i. Therefore, the first part—that is, $\sum_{V_1} \cdots \sum_{V_N}$—is nested sums over all inheritance patterns at all loci under consideration. For two full siblings, the length of each V_i is 16. As the Lander–Green algorithm starts from one side of the chromosome, the next part of the formula, $\Pr(V_1)$, describes the prior probability of the inheritance vector for the first marker, simply given by $1/\text{length}(V_1)$.

The next part, $\prod_{i=2}^{N} \Pr(V_i \mid V_{i-1})$, represents the transition probabilities between the hidden states in the Markov chain. (Notice that the product goes from marker 2 to marker N, since we start at marker 1, which is independent of previous loci by definition.) We compute the probability of the current inheritance vector at marker i, given a specific inheritance vector at the previous marker, $i - 1$. This probability usually evaluates to a product of ρ and $1 - \rho$ indicating the number of recombinations between the two inheritance vectors. For instance, given the vector [0 0 0 0] for marker $i - 1$ and the vector [0 0 0 1] for marker i, $\Pr(V_i \mid V_{i-1}) = \rho \times (1 - \rho)^3$. Comparing this with the notation presented in Example 4.2, we see that the inheritance vectors actually keep track of whether it is a male or a female transmission, and as noted in for instance [81], there appear to be considerable differences between female and male recombination maps.

The last part of the algorithm, $\prod_{i=1}^{N} \Pr(G_i \mid V_i)$, describes the product of probabilities of observing the specific genotypes, G_i, at marker i given the specific inheritance vector under consideration, V_i. Each probability typically evaluates to a product of allele frequencies.

Example 4.3 Application of the Lander–Green algorithm. We return to the example of disputed siblings, presented in Example 4.2, now instead applying the Lander–Green algorithm to derive the likelihood $\Pr(\text{data} \mid H_1)$; see Figure 4.7.

We start by defining the inheritance space for each marker, V_{L_1} and V_{L_2}. Disregarding the genotype data, we have 16 possible inheritance patterns, defined by the binary vector of length 4 (denoted by $v_{L_n,i}$ for marker n and pattern i); see Table 4.2.

As indicated in Figure 4.7, each individual meiosis is denoted by $v_{..k}$. For the first marker, L_1, we have that one or zero alleles may be IBD, and thus the number of possible inheritance vectors is reduced to 12. For instance, the vectors [0 0 0 0] and [1 1 1 1] will have a posterior probability of 0 as they indicate two alleles IBD. For the second marker, L_2, we have that zero alleles must be IBD, and therefore the set of inheritance vectors is reduced to 4. Using Equation 4.4 and the notation described above, we get

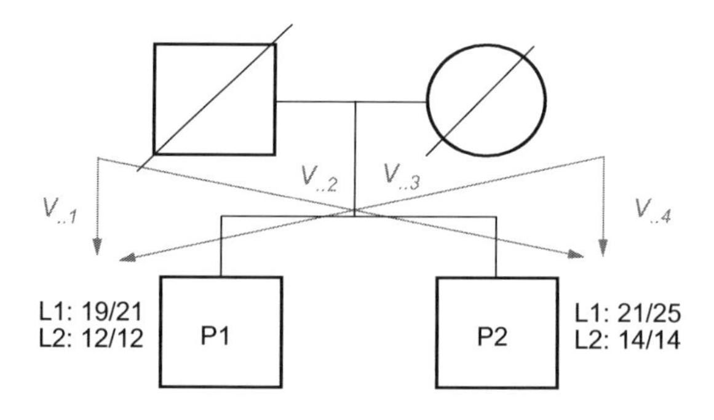

FIGURE 4.7

The inheritance patterns, denoted $v_{..1}, v_{..2}, v_{..3}, v_{..4}$.

Table 4.2 Inheritance Patterns and Conditional Probabilities of Observing the Genotypes at Each Marker Given the Specific Inheritance Pattern

$v_{..1}$	$v_{..2}$	$v_{..3}$	$v_{..4}$	$v_{..}$	$\Pr(G_{L_1} \mid v_{..})$	$\Pr(G_{L_2} \mid v_{..})$
0	0	0	0	[0 0 0 0]	0	0
0	0	0	1	[0 0 0 1]	$p_{19}p_{21}p_{25}$	0
0	0	1	0	[0 0 1 0]	$p_{19}p_{21}p_{25}$	0
0	0	1	1	[0 0 1 1]	$4p_{19}p_{21}^2p_{25}$	$p_{12}^2p_{14}^2$
0	1	0	0	[0 1 0 0]	$p_{19}p_{21}p_{25}$	0
0	1	0	1	[0 1 0 1]	0	0
0	1	1	0	[0 1 1 0]	$4p_{19}p_{21}^2p_{25}$	$p_{12}^2p_{14}^2$
0	1	1	1	[0 1 1 1]	$p_{19}p_{21}p_{25}$	0
1	0	0	0	[1 0 0 0]	$p_{19}p_{21}p_{25}$	0
1	0	0	1	[1 0 0 1]	$4p_{19}p_{21}^2p_{25}$	$p_{12}^2p_{14}^2$
1	0	1	0	[1 0 1 0]	0	0
1	0	1	1	[1 0 1 1]	$p_{19}p_{21}p_{25}$	0
1	1	0	0	[1 1 0 0]	$4p_{19}p_{21}^2p_{25}$	$p_{12}^2p_{14}^2$
1	1	0	1	[1 1 0 1]	$p_{19}p_{21}p_{25}$	0
1	1	1	0	[1 1 1 0]	$p_{19}p_{21}p_{25}$	0
1	1	1	1	[1 1 1 1]	0	0

Notes: *1 indicates that the paternal allele has been transmitted, while 0 indicates that the maternal allele has been transmitted.*

$$L(H_1) = \sum_{i=1}^{14}\sum_{j=1}^{4} \Pr(v_{L_1,i})\Pr(v_{L_2,j} \mid v_{L_1,i})\Pr(G_{L_1} \mid v_{L_1,i})\Pr(G_{L_2} \mid v_{L_2,j}),$$

where $v_{L_1,i}$ and $v_{L_2,j}$ indicate a specific inheritance vector at L_1 and L_2, respectively. Indeed the equation above is not easily evaluated by hand as we have 14×4 different inheritance combinations to sum over, but is rather calculated with use of a computer. In fact a simple R script (see Chapter 5) may be created to check the formula.

For illustrative purposes, we can evaluate the first term in the nested sum where $i = 1$ and $j = 1$, equal to the two inheritance vectors $v_{L_1,1} = [0\ 0\ 0\ 1]$ and $v_{L_2,1} = [0\ 0\ 1\ 1]$:

$$\begin{aligned}&\Pr(v_{L_1,1})\Pr(v_{L_2,1} \mid v_{L_1,1})\Pr(G_{L_1} \mid v_{L_1,1})\Pr(G_{L_2} \mid v_{L_2,1})\\ &= 1/16 \times \rho(1-\rho)^3 \times p_{19}p_{21}p_{25} \times p_{12}^2p_{14}^2.\end{aligned}$$

4.1.4 EXTENSIONS

The Lander–Green algorithm may be extended to accommodate other effects. The following sections give a brief explanation of how these extensions could be implemented, without providing detailed algorithms, and discuss the feasibility of those implementations. A simple notation will be used and explained where necessary.

X-chromosomal markers

To extend the Lander–Green algorithm to accommodate X-chromosomal markers, we need make only minor changes. For males, only one allele can be transmitted, and thus the inheritance vector for male meioses will contain only a single value, which is equal to the value for the meiosis resulting in his genotype. This is derived biologically from the fact that males possess only one X chromosome and practically no recombination occurs during the meiosis. As a consequence, likelihood calculations on X-chromosomal marker data may, for some pedigrees, be less computer intensive compared with likelihood calculations on autosomal markers. For instance, the inheritance space for two full brothers only contains two elements indicating the maternal meioses, whereas for autosomal markers the inheritance vectors are defined by four elements indicating maternal and paternal meioses.

Mutations

In forensics, mutations are of great importance as genetic inconsistencies are commonly encountered. This is particularly true for STR markers with high mutation rates. Proper mutation models are important not only for paternity/maternity cases but also for more complex pedigrees, such as, for instance, the case of three full siblings where mutations cannot be directly observed, but may be necessary in order to explain the data.

The Lander–Green algorithm does not easily accommodate mutations as we do not keep track of allele transmission throughout the pedigree (remember that inheritance vectors keep track only of the meioses). Instead the algorithm considers IBD probabilities conditional on identical-by-state states. With mutations two alleles may, in a sense, be IBD even though they are not identical by state. Consider, for instance, two full siblings, both being homozygous for different alleles, 12/12 and 14/14. The likelihood for this single marker can then be formed as

$$L = \Pr(\text{data} \mid \text{full siblings}) = \sum_{v_i} \Pr(v_i) \times \Pr(G \mid v_i), \tag{4.5}$$

where the v_i indicate inheritance vectors, in the current example consisting of four elements indicating whether each allele for the siblings has been inherited from the paternal or the maternal chromosome of the parents. The last probability, $\Pr(G \mid v_i)$, is where mutations would typically be accounted for as several of these probabilities would otherwise be 0; see Table 4.2. Consider the inheritance vector [1 1 1 1] implying that the two siblings have inherited their alleles from the same parental and maternal alleles and that they share exactly two alleles IBD. For mutations, $\Pr(G \mid v_i)$ would then be formed as a summation over the possible parental alleles leading to the genotypes and would be a product of transition probabilities and allele frequencies.

One extension of the Lander–Green algorithm was proposed by Kling et al. [85], where founder alleles are used to keep track of specific allele transmissions in the pedigree in addition to accounting for linkage. Using founder allele patterns, we can easily account for mutations. While this approach works well in theory, the complexity grows fast with the number of transmissions in the pedigree. See also Section 4.2.2 for a detailed description.

Subpopulation correction

To allow for subpopulation structures we must adjust the allele frequencies using θ correction; see also Equation 2.13. The accommodation of subpopulation correction in the Lander–Green algorithm is in fact quite straightforward. Through the inheritance vectors we keep track of the number of founder alleles in the pedigree—that is, we know how many alleles we need to sample for each inheritance vector. We return to the simple example in the previous section, but now consider two alleged full siblings, both being homozygous 12/12. We may use a condensed variant of the Lander–Green algorithm, where we consider only the IBD states 0, 1, and 2 in a comparison to complete summation over individual inheritance vectors as in the previous section. We have

$$L = \Pr(\text{data} \mid \text{full siblings}) = 0.25g_0 + 0.5g_1 + 0.25g_2, \tag{4.6}$$

where $g_i = \Pr(\text{data} \mid \text{IBD} = i)$. Disregarding mutations, we get $g_2 = p_{12}^2$ and $g_1 = p_{12}^3$, while $g_0 = p_{12}^4$. In other words, for each g_i we sample $4 - i$ allele 12, and the sampling formula, described in Equation 2.13, could easily be adopted to adjust the allele frequencies accordingly. This extension can be implemented without special considerations to linkage between markers as $\Pr(G_i|V_i)$ in the Lander–Green algorithm (corresponding to g_0, g_1, and g_2 in the formulation above) is independently computed for each marker.

Dropouts and silent alleles

We may further consider extensions of the Lander–Green algorithm to account for allelic dropouts (described in Section 2.7) and silent alleles (described in Section 2.6). Dropouts would typically be accommodated for by the same methods as presented in [46] for unlinked markers. We would have to sum over the possible genotype constellations for a given marker and compute the likelihood for each combination. A similar approach would have to be considered for silent alleles. These concepts would require considerable computational effort in some cases—for example, where the genotype data contain a great number of homozygotes. We do not cover the computational details here.

4.2 LINKAGE DISEQUILIBRIUM

In contrast to linkage, linkage disequilibrium (LD) occurs in populations and is not explicitly biological, even though it originates in biological phenomena. Disequilibrium occurs when alleles at different loci appear together at rates that differ from what would be expected under independence. The nonrandom association of alleles may be caused by close proximity of markers on the same chromosome, but also by phenomena such as nonrandom mating, mutations, and population substructures.

It is important to understand the distinction between linkage, occurring within a pedigree, causing a specific haplotype to be inherited as a unit, without recombination, and LD, which is observed in a population as the increased/decreased occurrence of certain haplotypes, which may, as previously mentioned, be a consequence of low recombination. The latter causes certain haplotypes to be accumulated and leads to disequilibrium for the alleles; see also Figure 4.1 for an illustration of the distinction between linkage and LD.

Example 4.4 Calculating a measure of LD. LD is more specifically measured as the difference between expected and observed haplotype frequencies. To illustrate this, consider two diallelic loci S_1 and S_2 with alleles A, a and B, b, respectively. Table 4.3 contains the number of observations for each combination of alleles at the two loci. It is easy to derive the allele frequencies as $p_A = 40/100$, $p_a = 60/100$, $p_B = 60/100$, and $p_b = 40/100$. We may further compute the expected haplotype frequencies by merely multiplying the allele frequencies, and so for haplotype $[A\ B]$ the expected frequency is $p_A \times p_B = 0.24$. The observed haplotype frequencies may also be derived from Table 4.3; for the haplotype mentioned we get $H_{A,B} = 5/100$. A simple measure of the LD is computed as

$$D = H_{A,B} - p_A p_B = -0.19.$$

For diallelic loci where $p_a = (1 - p_A)$, it can be demonstrated that $|D|$ is the same for all four haplotype combinations—for example,

$$|D| = |H_{A,B} - p_A p_B| = |H_{a,b} - p_a p_b| = 0.19.$$

A commoner measure of LD is the squared correlation:

$$r^2 = \frac{(H_{A,B} - p_A p_B)^2}{p_A p_a p_B p_b},$$

which is a normalized version of D.

The above discussion may be expanded to include markers with any number of alleles. Consider a matrix Δ with elements d_{ij}, corresponding to all combinations of alleles at two loci, $i = 1, \ldots, I$, where I is the total number of alleles at the first locus. Similarly $j = 1, \ldots, J$, where J is the total number of alleles at the second locus. Then, $d_{i,j}$ indicates the disequilibrium between alleles i and j.

Table 4.3 Contingency Table for the Haplotypes

	A	a
B	5	55
b	35	5

LD will disappear over time due to recombination, which can be demonstrated with the elements of Δ, as introduced above. If we do not consider mutations and other effects, we have

$$d_{ij} = H_{i,j} - p_i p_j \tag{4.7}$$

for each element of the matrix Δ. Given one meiosis—that is, a new generation—we can compute the new disequilibrium according to

$$d'_{ij} = \left(H_{i,j}(1-\rho) + p_i p_j \rho\right) - p_i p_j, \tag{4.8}$$

where d'_{ij} denotes the disequilibrium in the second generation. The first part of the right-hand side of the equation describes the number of haplotypes that have not recombined, while the second part describes the number of haplotypes that have recombined. The allele frequencies are assumed to be constant. Rewriting the equation, we get the recursive formula

$$d'_{ij} = (H_{i,j} - p_i p_j) - \rho(H_{i,j} - p_i p_j) = d_{ij} - \rho d_{ij} = (1-\rho)d_{ij},$$

where the degree of LD in the second generation, d'_{ij}, has decreased with $(1-\rho)$. It follows from this that $d^n_{ij} = (1-\rho)^n d_{ij}$ and $\lim_{n\to\infty} d^n_{ij} = 0$. The decay of LD with increasing number of generations for some values of ρ is shown in Figure 4.8.

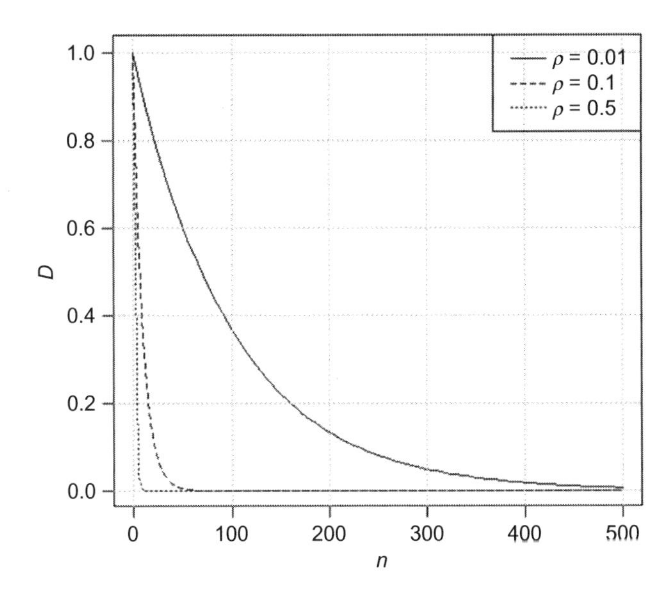

FIGURE 4.8

The decay of LD illustrated by the variable D as a function of the number of generations n for some values of ρ.

It is apparent that only a few generations are required if the loci are unlinked—that is, if $\rho = 0.5$—while if the loci are tightly linked—for instance, $\rho = 0.01$—a great number of generations, more than 500, is required to break down the LD completely. *Note*: These derivations disregard other effects that could increase the degree of LD.

4.2.1 INTRODUCTION TO CALCULATIONS

In likelihood calculations accounting for LD, we must consider haplotypes instead of genotypes. The main obstacle is the fact the haplotypes may not be unambiguously determined. For instance, consider the two full siblings in Figure 4.9, where we see that the haplotype for the individual denoted P1 can be determined because he is homozygous for one locus. In contrast, for P2 we see that two different haplotype setups may exist, and as a consequence we have to consider each possibility.

For a reason that will become obvious in the following sections, we introduce another concept—conditional allele probability—that will be used in some theoretical derivations. As we shall see, this probability is closely intertwined with the haplotype frequency. Mathematically, the conditional probability $\Pr(x \mid y)$, indicating the probability of observing allele x for one locus given that allele y has been observed at a second locus, is found through the relation $H_{x,y} = \Pr(x \mid y)p_x$, where $H_{x,y}$ is the frequency of haplotype $[x\ y]$ and p_x is the frequency of allele x. In summary, instead of considering the loci as a unit, with haplotypes, we may treat them separately and model the dependency between the alleles using conditional allele probabilities and the dependency between the markers using methods as described in Section 4.1.3.

Example 4.5 Maternity case with LD. We start by considering a simple maternity case. Here, simple in the sense that we have only a direct relationship. Even so, the phase of the haplotypes may not be certain. Consider two linked diallelic SNP markers, S_1 and S_2, where we know association exists between the alleles. The hypotheses are as follows:

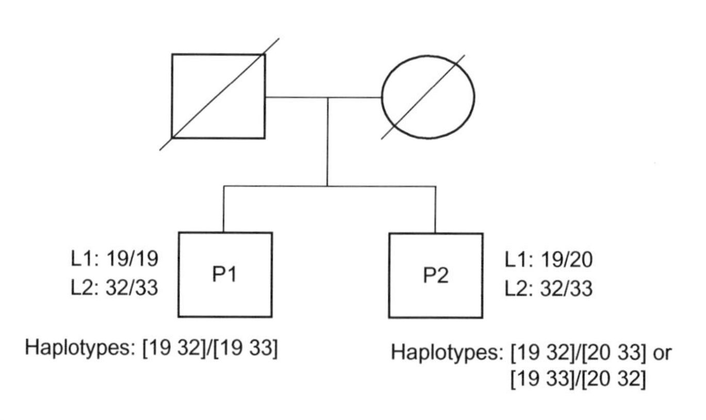

FIGURE 4.9

Problems to determine the exact phase for haplotypes shown for two full siblings.

Table 4.4 Genotype Data for the Example in Section 4.5

	Alleged Mother	Child
M_1	A/a	a/a
M_2	B/b	b/b

H_1: The alleged mother is the true mother of a female child.
H_2: The alleged mother and the child are unrelated.

Furthermore, assume we have data as indicated in Table 4.4. Starting with the likelihood given H_1, we get

$$\begin{aligned}\Pr(\text{data} \mid H_1) &= H_{a,b}\left(0.5\rho \times 2H_{A,b}H_{a,B} + 0.5(1-\rho) \times 2H_{A,B}H_{a,b}\right)\\ &= /\text{switching to conditional allele probability notation}/\\ &= p_A p_a^2 \Pr(b \mid a)\,(\rho \Pr(b \mid A)\Pr(B \mid a)\\ &\quad +(1-\rho)\Pr(B \mid A)\Pr(b \mid a)),\end{aligned}$$

where we consider both haplotype setups for the mother, and the transmission probability is given as $0.5(1-\rho)$ when a recombination is not required and 0.5ρ otherwise. Similarly, the likelihood given H_2 is given by

$$\begin{aligned}\Pr(\text{data} \mid H_2) &= H_{a,b}^2(2H_{A,b}H_{a,B} + 2H_{A,B}H_{a,b})\\ &= p_a^2 \Pr(b \mid a)^2 \times 2p_A p_a(\Pr(b \mid A)\Pr(B \mid a) + \Pr(B \mid A)\Pr(b \mid a)),\end{aligned}$$

and finally

$$\begin{aligned}\text{LR} &= \frac{\Pr(\text{data} \mid H_1)}{\Pr(\text{data} \mid H_2)}\\ &= \frac{\rho\Pr(b \mid A)\Pr(B \mid a) + (1-\rho)\Pr(B \mid A)\Pr(b \mid a)}{2p_a \Pr(b \mid a)(\Pr(b \mid A)\Pr(B \mid a) + \Pr(B \mid A)\Pr(b \mid a))}.\end{aligned}$$

Using $p_A = 0.8$, $p_a = 0.2$, $p_B = 0.7$, and $p_b = 0.3$, we may plot the LR as a function of the disequilibrium $|D|$ for some values of ρ; see Figure 4.10. *Note*: We need two different plots as the sign of D may be either positive or negative, while $|D|$ is always positive. Regardless of the sign of D, the LR will approach $1/(4p_a p_b)$ as D approaches zero. Otherwise, the plots are completely different; if D is negative, the LR will increase and approach infinity as $|D|$ approaches 0.06 (the maximum value), while if D is positive, the LR will decrease as D approaches 0.14, and it can be shown that the LR will approach $(1-\rho)/\,(2(|D| + p_a p_b))$.

4.2.2 *GENERALIZATION

To formulate a general algorithm for likelihood calculations where models for LD structure are included is more complex and usually requires greater computational

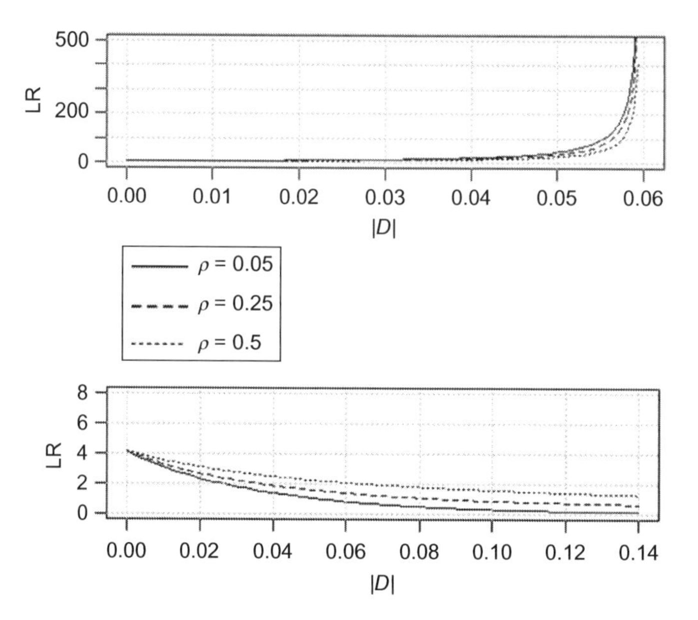

FIGURE 4.10

The LR in a maternity case as a function of the LD ($|D|$) for some values of the recombination rate (ρ). There are two different plots as the true value of D can be either positive (top) or negative (bottom).

effort as a consequence of the uncertainty about haplotype phases. There are, however, some simplifications that may be assumed in order for feasible derivations to be made.

Cluster approach

One approach is to consider clusters of markers, where recombination within a cluster is disregarded. Briefly, for each genotyped individual we iterate over all possible haplotypes in a cluster, disregarding mutations; see Figure 4.11.

Conditional on the phase of the haplotypes we use the Lander–Green algorithm to efficiently compute the likelihood. As we assume complete linkage within a cluster, the haplotypes, once instantiated, can be compared with single marker alleles that cannot change in the pedigree, unless we consider mutations. The approach will essentially be a summation over all possible haplotype setups for all typed individuals.

As we do not consider recombinations, calculations may be swift but may also cause the likelihood to become zero; see Figure 4.12, where an obligate recombination is necessary in order to explain the data.

Details on an efficient implementation of the cluster approach are available in Abecasis et al. [86].

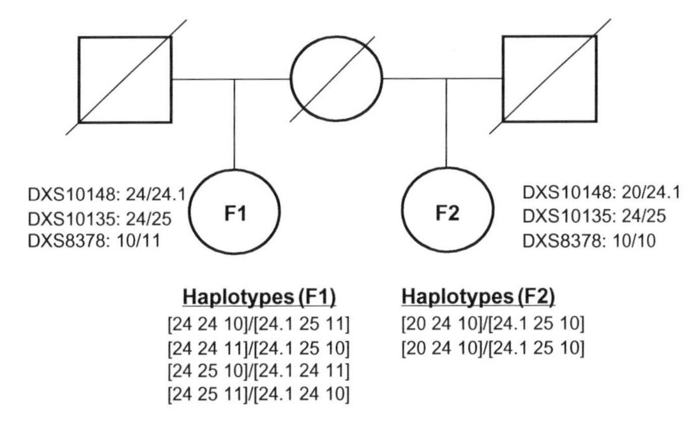

FIGURE 4.11

Listing the different haplotype setups for two females, denoted F1 and F2 in a cluster of three X-chromosomal STR markers.

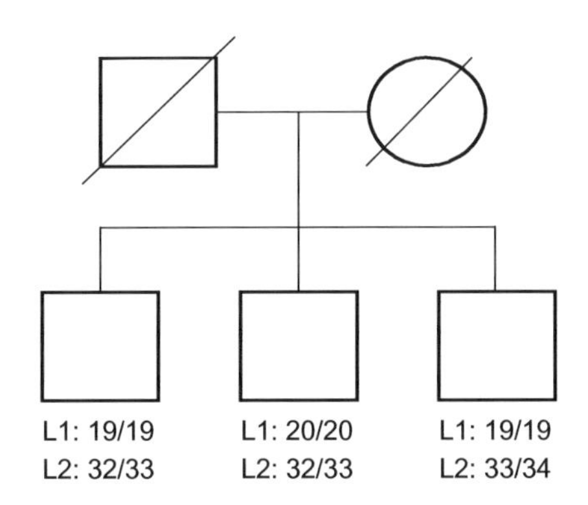

FIGURE 4.12

Obligate recombination within a cluster of two completely linked autosomal markers L_1 and L_2. The case depicts a case of three full siblings where five different haplotypes are observable, while the parents can maximally have four different haplotypes.

Example 4.6 Illustration of the cluster approach. Consider the example of two cousins (see Figure 4.13) with genotype data as indicated in Table 4.5.

To obtain the likelihood for the hypotheses of the cousin relationship, we sum over all possible haplotype setups for the two females:

$$L = \sum_{S_1} \left(\sum_{V_1} \Pr(V_1)\Pr(G_1 \mid V_1, S_1) \right),$$

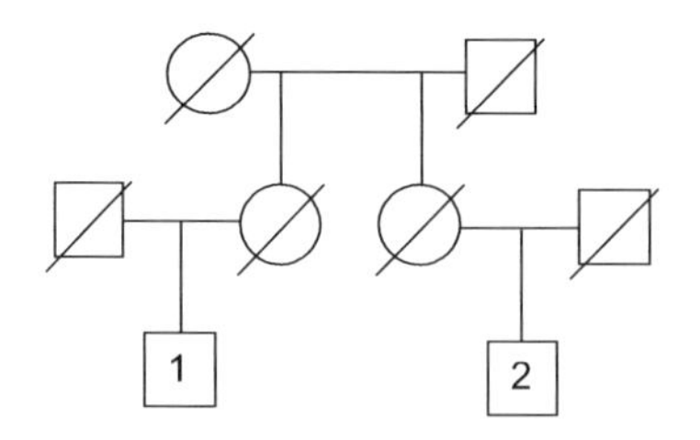

FIGURE 4.13

Two cousins.

Table 4.5 Genotype Data for Cousins 1 and 2 for Two Genetic Markers M_1 and M_2

	1	2
M_1	A/a	A/a
M_2	B/b	B/b

where the outer sum is over possible haplotype setups (phases), S_1 for cluster 1. As we consider only one cluster of markers, we do not need to consider recombinations between clusters. V_1 denotes the inheritance space for the cluster—that is, in contrast to the Lander–Green algorithm, we consider inheritance patterns for clusters and not for markers. $\Pr(G_1 \mid V_1, S_1)$ is the conditional probability of observing the genotypes given an inheritance pattern V_1 and a specific phase S_1.

In fact, we may in our case instead use the plain IBD formula for pairs of individuals (see Section 2.6) to evaluate the likelihood. Rewriting the formula above, we get

$$\begin{aligned} L &= \sum_{S_1} \left(\sum_{V_1} \Pr(V_1)\Pr(D_1 \mid V_1, S_1) \right) \\ &= \sum_{S_1} \left(\sum_{i=0}^{2} \Pr(\text{IBD} = i) \Pr(\text{data} \mid \text{IBD} = i, S_1) \right). \end{aligned}$$

In total, there are two possible haplotype setups for each person, and thus the size of S_1 is 4, given by the combination of all setups. Using $H_{A,B} = 0.2$, $H_{A,b} = 0.1$, $H_{a,B} = 0.6$, and $H_{a,b} = 0.1$ and for two cousins $\Pr(\text{IBD} = 0) = 0.75$, $\Pr(\text{IBD} = 1) = 0.25$, and $\Pr(\text{IBD} = 2) = 0$, we get

Table 4.6 Results From a Calculations Using the Cluster Approach. Last Line Sums Probabilities

Haplotypes for 1	Haplotypes for 2	Pr(data | H_1, S_1)	Pr(data | H_2, S_1)
[A B]/[a b]	[A B]/[a b]	0.0015	0.0016
[A B]/[a b]	[A b]/[a B]	0.0036	0.0048
[A b]/[a B]	[A B]/[a b]	0.0036	0.0048
[A b]/[a B]	[A b]/[a B]	0.0011	0.0144
		0.0098	0.0256

$$
\begin{aligned}
L &= \sum_{S_1}\left(\sum_{i=0}^{2} \Pr(\text{IBD}=i)\Pr(\text{data} \mid \text{IBD}=i, S_1)\right)\\
&= \left(0.75 \times (2H_{A,B}H_{a,b})^2 + 0.25 \times (H_{A,B}H_{a,b}(H_{A,B}+H_{a,b}))\right)\\
&\quad + \cdots + \left(0.75 \times (2H_{A,b}H_{a,B})^2 + 0.25 \times (H_{A,b}H_{a,B}(H_{A,b}+H_{a,B}))\right),
\end{aligned}
$$

which is also illustrated in Table 4.6, where H_2 is the hypothesis indicating that the two individuals are unrelated and where $\Pr(\text{data} \mid H_k, S_1) = \sum_{i=0}^{2} \Pr(\text{IBD} = i)$ $\Pr(\text{data} \mid \text{IBD} = i, S_1))$ for hypothesis k.

The LR is finally computed as 0.0098/0.0256 = 0.38, which may be confirmed in `Merlin` [90]. In other words, even though the two males share the same genotypes, the uncertainty in haplotype phase in combination with the LD structure favors the alternative hypothesis of unrelatedness.

The cluster approach is a sufficient approximation under some circumstances and is easy to implement. We next look at an alternative approach where some of the simplifications assumed in the cluster approach are no longer necessary.

Exact calculations

An alternative approach, as introduced by Kling et al. [85], is to consider founder alleles patterns; see also Kurbasic et al. [87]. In addition to building inheritance graphs, we can set up founder allele patterns—that is, specify the founders of a pedigree—and let their genotypes act as the set of alleles we are interested in; see Figure 4.14.

Using condensed notation, we may summarize the algorithm as

$$
\begin{aligned}
\Pr(\text{data} \mid H) &= \sum_{V_1}\cdots\sum_{V_N}\sum_{F_1}\cdots\sum_{F_N}\Pr(V_1)\Pr(F_1)\\
&\prod_{i=2}^{N}\Pr(V_i \mid V_{i-1})\prod_{i=2}^{N}\Pr(F_i \mid F_{i-L},\ldots,F_{i-1})\prod_{i=1}^{N}\Pr(D_i \mid V_i, F_i),
\end{aligned}
\tag{4.9}
$$

where we in addition to summing over the inheritance space now also sum of all founder allele patterns. The probability $\Pr(F_i \mid F_{i-L},\ldots,F_{i-1})$ is usually a product

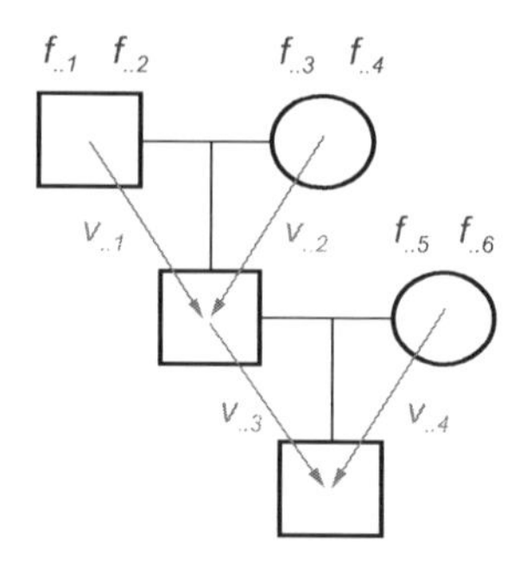

FIGURE 4.14

A pedigree where all meioses are listed, $v_{..1}, \ldots, v_{..4}$, as well as founder alleles, $f_{..1}, \ldots, f_{..6}$.

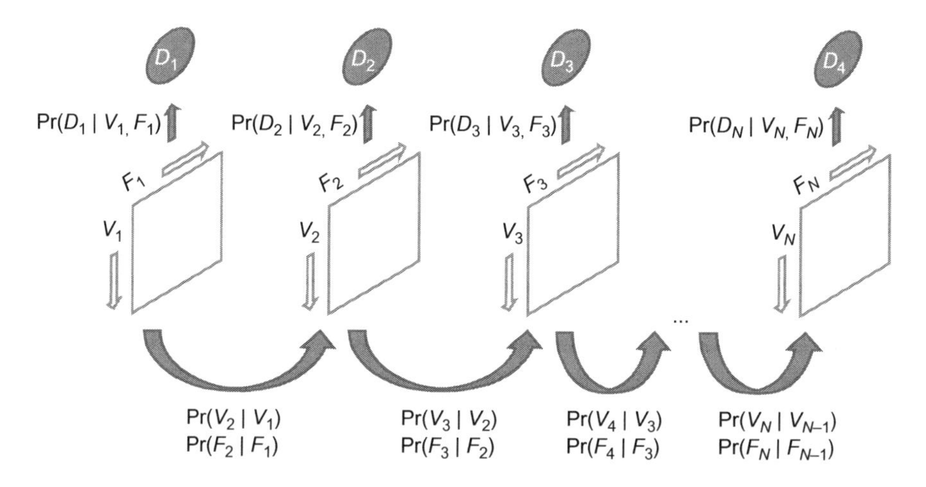

FIGURE 4.15

Illustration of the exact algorithm as discussed in Section 4.2.2, here depicted with use of a one-step Markov chain for founder allele patterns.

of the conditional probabilities for the founder alleles at marker i given the founder alleles at markers $i-L, \ldots, i-1$, where L indicates the number of markers for which we model LD structure. The model uses Markov chains as illustrated in Figure 4.15.

Example 4.7 Illustration of the exact approach. Consider two sisters (F1 and F2) with data available from their common mother (see Figure 4.16), where the hypothesis is that they also share a common father.

We assume data for two X-chromosomal markers (X1 and X2) as indicated in the figure. Using Equation 4.9, we can now derive the theoretical formula for the likelihood as (disregarding mutations)

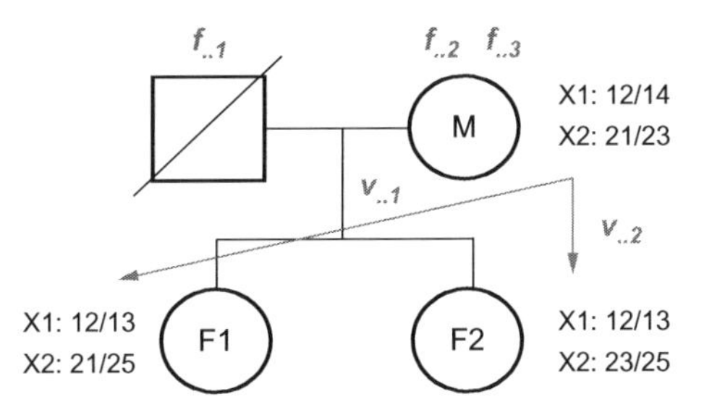

FIGURE 4.16

Two full sisters, denoted F1 and F2, with genotype data for their mother, M.

$$\Pr(\text{data} \mid H = \text{full siblings}) = \sum_{V_1}\sum_{V_2}\sum_{F_1}\sum_{F_2} \Pr(V_1)\Pr(F_1)\Pr(V_2 \mid V_1)$$
$$\Pr(F_2 \mid F_1)\Pr(D_1 \mid V_1, F_1)\Pr(D_2 \mid V_2, F_2).$$

As we consider X-chromosomal data, there are only two meioses of interest, indicating whether each sister has inherited the paternal (1) or the maternal (0) allele of the mother. Therefore, the length of V_1 and V_2 is 4, defined by the set {[0 0], [0 1], [1 0], [1 1]}. *Note*: Some patterns may indicate states where the likelihood is zero (unless mutations are accounted for). For instance, at X1 only states [0 0] and [1 1] are possible as the sisters must have inherited the same allele (12) from the mother. Similarly, for X2 only states [0 1] and [1 0] are possible. The transition probabilities between the states, $\Pr(V_2 \mid V_1)$, are given in Table 4.7.

The founder allele patterns, denoted F_1 and F_2 for the two markers, respectively, are defined by the two alleles of the mother and the possible alleles for the father. The latter in this case is restricted to allele 13 for X1 and allele 25 for X2. For instance, the founder allele set for F_1 is defined by {[12 14 13], [14 12 13]}. Each element of the set is specified such that the first and second elements represent the maternal and paternal alleles of the mother respectively. *Note*: If mutation were accounted for, other possible founder alleles would be possible for the

Table 4.7 Transition Probabilities Between Inheritance States (V_1 and V_2) at Two Markers

V_1	V_2	$\Pr(V_2 \mid V_1)$
[0 0]	[0 1]	$(1-\rho)\rho$
[0 0]	[1 0]	$\rho(1-\rho)$
[1 1]	[0 1]	$\rho(1-\rho)$
[1 1]	[1 0]	$(1-\rho)\rho$

Note: *ρ is the recombination rate.*

untyped father. The probability $\Pr(F_1)$ is simply a product of allele frequencies—for instance, $\Pr([12\ 14\ 13]) = p_{12}p_{14}p_{13}$. The transition probabilities between the states, $\Pr(F_2 \mid F_1)$, are given in Table 4.8, and are typically a product of conditional allele probabilities.

Finally, the terms $\Pr(D_1 \mid V_1, F_1)$ and $\Pr(D_2 \mid V_2, F_2)$ are computed as given in Table 4.9. The probabilities assume either the value 0 or the value 1 (as we disregard mutations) indicating if the data are consistent with the inheritance pattern and the founder allele pattern. *Note*: Table 4.9 does not contain an exhaustive list of all combinations for V_1, V_2, F_1, and F_2, but includes all the necessary probabilities needed in the summation.

Combining the information in the tables and rearranging the formula, we get

$$\Pr(\text{data} \mid H = \text{full siblings}) = 1/4\rho(1-\rho)p_{13}p_{12}p_{14}\Pr(25 \mid 13)(2\Pr(21 \mid 12)\Pr(23 \mid 14) + 2\Pr(21 \mid 14)\Pr(23 \mid 12)).$$

Using haplotype data as given in Table 4.10 and $\rho = 0.01$, we get $\Pr(\text{data} \mid H = \text{full siblings}) = 1.12 \times 10^{-6}$

Table 4.8 Transition Probabilities Between Founder Allele States (F_1 and F_2) at Two Markers

F_1	F_2	$\Pr(F_2 \mid F_1)$
[12 14 13]	[21 23 25]	Pr(21 | 12) Pr(23 | 14) Pr(25 | 13)
[12 14 13]	[23 21 25]	Pr(23 | 12) Pr(21 | 14) Pr(25 | 13)
[14 12 13]	[21 23 25]	Pr(21 | 14) Pr(23 | 12) Pr(25 | 13)
[14 12 13]	[23 21 25]	Pr(23 | 14) Pr(21 | 12) Pr(25 | 13)

Table 4.9 Combinations of Inheritance Patterns (V_1 and V_2) and Founder Allele Patterns (F_1 and F_2)

V_1	V_2	F_1	F_2	$\Pr(D_1 \mid V_1, F_1)$	$\Pr(D_2 \mid V_2, F_2)$
[0 0]	[0 1]	[12 14 13]	[21 23 25]	1	1
[0 0]	[1 0]	[12 14 13]	[23 21 25]	1	1
[1 1]	[0 1]	[14 12 13]	[21 23 25]	1	1
[1 1]	[1 0]	[14 12 13]	[23 21 25]	1	1

Table 4.10 Haplotype Observations

	12	13	14
21	585	0	10
23	10	0	385
25	0	10	0

Similar derivations may be applied to the alternative hypotheses that the two individuals are half sisters, with different fathers. Omitting the details, we get

$$\Pr(\text{data} \mid H = \text{half siblings}) = 1/4\rho(1-\rho)p_{13}^2 p_{12} p_{14} \Pr(25 \mid 13)^2 (2\Pr(21 \mid 12)\Pr(23 \mid 14) + 2\Pr(21 \mid 14)\Pr(23 \mid 12)),$$

and finally the LR comparing the two hypotheses becomes 100 in favor of full siblings. The observant reader will notice that the final LR is independent of the recombination rate ρ as all terms containing ρ cancel out.

Keeping track of founder alleles in addition to using inheritance vectors, we may actually include feasible models for mutations. The drawback with this more exact approach is the computational effort, which quickly grows with the number of alleles for each marker. However, for diallelic SNP markers, where the number of possible founder allele patterns is small, the algorithm is comparatively swift.

4.3 HAPLOTYPE FREQUENCY ESTIMATION

All computational models accounting for LD described thus far necessitate estimates of haplotype frequencies. Although methods exist to provide such estimates, there is still a large uncertainty. This holds particularly for STR markers, where the number of possible haplotypes grows rapidly with the number of alleles for each marker considered. For instance, consider a cluster of M genetic markers where, for simplicity, each marker has A alleles; the number of possible haplotypes then equals A^M. For a cluster with three STR markers where the number of alleles is 15, which is not unrealistic, the number of possible haplotypes would be 3375.

For comparison, we may consider a simple scenario. STR markers commonly have, say 10-30 different "possible" alleles, and studies on population allele frequencies include in the range of 100-1000 samples, at least 10 times the number of possible variants. If the sample size required to obtain good estimates for haplotype frequencies were as demanding, at least 30,000 individuals would be required in the above example with clusters of three markers. *Note*: These derivations assume a simplified model to estimate frequencies, see Section 6.1 for a more detailed discussion.

The point with this analogy is that, in general, greater number of samples (sampled individuals) are required when collecting data for haplotype databases compared with allele databases.

We now turn our focus to models for haplotype frequency estimation. Tillmar et al. [88] introduced the so-called λ model (see also Sections 6.1.4 and 7.3)

$$F_i = \frac{c_i + \lambda p_i}{C + \lambda} \tag{4.10}$$

relating the estimated haplotype frequency F_i for a specific haplotype i to the observations denoted by c_i for the counts. We have

$$\sum_{i=1}^{N} c_i = C.$$

In other words, C is the total number of observations in the haplotype database with N different haplotypes.

To obtain frequencies also for unobserved haplotypes, the model introduces the parameter λ, giving weight to the expected frequency p_i of a haplotype. Indeed, the λ model requires as input only a table of observations; the expected frequencies can be derived directly from these.

Example 4.8 Illustrating the λ model. We use small clusters of two SNP markers to provide an illustration of the λ model. Consider the haplotype data given in Table 4.11. Using the table, we may derive the allele frequencies as $p_A = 0.5$, $p_a = 0.5$, $p_B = 0.1$, and $p_b = 0.9$. We further have $C = 200$. An illustration of the resulting haplotype frequency estimates obtained with the λ model is given in Table 4.12 for some values of λ. The expected haplotype frequency, p_i, is computed as, for instance, $0.5 \times 0.1 = 0.05$ for the haplotype $[A\ B]$. We note that as λ assumes increasing values, the estimated frequencies, F_i, converge with the expected frequencies, p_i. Conversely, when λ approaches zero, F_i converges with the observed haplotype frequencies, c_i/C.

Example 4.9 Illustrating the effect of the λ model. To better illustrate how haplotype frequency estimation with the λ model (Equation 4.10) affects the LR, we may use a simplified case. Assume two females, F1 and F2, share the same homozygous haplotype for three X-chromosomal STR markers. We consider the following:

H_1: F1 and F2 are paternal half siblings.
H_2: F1 and F2 are unrelated.

Table 4.11 Haplotype Observations for Two Markers

	A	*a*	
B	2	18	20
b	98	82	180
	100	100	

Table 4.12 Haplotype Frequency Estimates

Haplotype	p_i	c_i	F_i ($\lambda = 0$)	F_i ($\lambda = 1$)	F_i ($\lambda = 200$)	F_i ($\lambda = 20{,}000$)
[*A B*]	0.05	2	0.01	0.01	0.03	0.05
[*A b*]	0.45	98	0.49	0.49	0.47	0.45
[*a B*]	0.05	18	0.09	0.09	0.07	0.05
[*a b*]	0.45	82	0.41	0.41	0.43	0.45

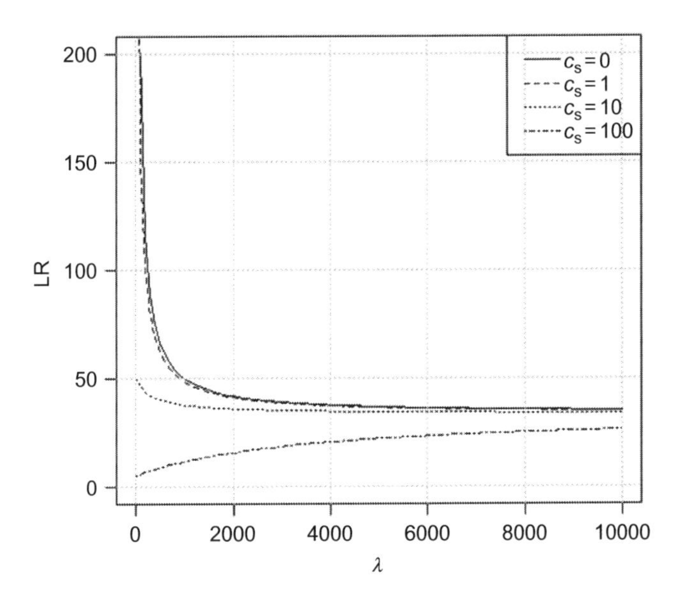

FIGURE 4.17

The dependence of the LR on λ for some values of c_S.

Given the situation where F1 and F2 share exactly the same haplotypes (all loci are homozygous), we may write (disregarding mutations)

$$\text{LR} \propto \frac{1}{F_S} = \frac{C + \lambda}{c_S + \lambda p_S}, \tag{4.11}$$

where F_S is frequency of the shared haplotype, c_S is the number of observations for this haplotype, and p_S is the expected frequency of the haplotype. The above formulation is simply an inversion of Equation 4.10. In other words, we may use

$$\text{LR} = \frac{C + \lambda}{c_S + \lambda p_S}$$

to illustrate the effect of the λ model on the results.

Typically, the value of p_S is small for STR markers, in the range from 0.3^M to 0.01^M (and even smaller), where M is the number of markers in a haplotype group. For data published in [7, 89] the largest p_i is 0.022 for clusters with three STR markers. Fixing $p_S = 0.03$ and $C = 500$ we can plot the LR as a function of λ for some values of c_S; see Figure 4.17, which illustrates the varying dependence of the LR on λ. Specifically, we see that for larger values of c_S—that is, the shared haplotype is common in the population—the LR does not depend as strongly on λ, whereas for lower values of c_S the dependence is strong. This is intuitive if we examine Equation

4.10 as if c_i is small (or zero) $\lambda \times p_i$ will dominate the frequency estimate, especially for large values of λ.

The λ model and its theoretical derivation will be more thoroughly explored in Chapter 6. The important point in this section was to illustrate the strong dependence between the LR and the parameters in the frequency estimation. This will be pursued further in the exercises.

4.4 PROGRAMS FOR LINKED MARKERS

Software for calculating likelihoods for linked markers is common in medical genetics, where linkage mapping is used to find links between specific genes and certain phenotypes. The webpage https://github.com/gaow/genetic-analysis-software lists a large number of programs alphabetically, and includes applications in genetic association analysis, haplotype construction, pedigree drawing, and population genetics. In forensic genetics, there was previously a shortage of good toolkits to analyze such data. Recent progress has provided two software programs where calculations with linked markers are possible, `FamLink` and `FamLinkX`. The following sections provide a brief description of the software, leaving complete details to [5, 85] as well as the exercises.

4.4.1 `FamLink`

`FamLink` was developed in response to the increasing focus on linked markers in the forensic community. `FamLink` calls the functions of `Merlin` [90] to compute likelihoods for postulated relationships. The core computations rely on the Lander–Green algorithm, described in detail in Section 4.1.3, and perform fast likelihood calculations with use of sparse gene flow trees; see Abecasis et al. [90] for details. `FamLink` furthermore includes the possibility to perform simulations in order to obtain some summary statistics for the distribution of the LR given different hypotheses about relationships. The latter is especially useful to study the effect of accounting/not accounting for linkage between two markers.

The main interface is currently restricted to two markers, whereas a special feature allows the analysis of complete `Familias` projects, then accounting for linkage between virtually any number of markers; see Exercise 4.6.

4.4.2 `FamLinkX`

`FamLinkX` was developed in response to the growing focus on STR markers on the X chromosome [85]. The core computations are described in [85], and are based on a version of the Lander–Green algorithm. Similarly to `FamLink`, `FamLinkX` is coded in the C++ language and provides an easy-to-use Windows interface, compatible with most Windows systems. `FamLinkX` computes likelihoods for hypotheses about pedigree structure given some X-chromosomal marker data. The algorithm embodies the idea of Markov chains, both for dependence across markers, similarly to the Lander–Green algorithm, and also for association of alleles across markers. The latter should not be confused with dependence of alleles at the same marker, usually modeled with subpopulation correction.

FamLinkX provides more extended and complex functionality than FamLink. For instance, an arbitrary number of markers may be defined. In contrast to other forensic software, FamLinkX defines clusters (also referred to as haplo/linkage groups) of markers. Within a cluster we define the markers and we specify haplotype observations. For X-chromosomal data, haplotype observations are most easily collected with data from males, as the phase cannot be unambiguously determined for females, and thus methods such as the expectation-maximum algorithm must be applied. The calculations are later performed with this haplotype information, but also accounting for the possibility of recombinations between markers within the cluster. The latter is sometimes necessary to explain the genotype data. In addition, recombination between markers at different clusters is accounted for. Similarly to Familias, FamLinkX also implements various models for mutations; see Kling et al. [11] for details.

4.5 EXERCISES

The following exercises introduce the computer programs FamLink (version 1.15) and FamLinkX (version 2.2) and how to deal with linked genetic markers. Software, input files and solutions (some in the form of videos) can be found following links from http://familias.name.

4.5.1 AUTOSOMAL MARKERS AND FamLink

Familias, used in previous chapters, does not consider linkage, so we introduce FamLink for pairs of linked autosomal markers. Some basic functionality will be explored as well as more advanced functions and theoretical points.

Note: The genders in the predefined figures may deviate from the ones in the exercise. This can, however, be overwritten in a subsequent step.

Exercise 4.1 Simple paternity case (video). We will first consider a simple exercise where the purpose is to get familiar with the user interface and create your frequency database. Normally a saved frequency database will be loaded. Consider a paternity case (duo) (for illustration, see Figure 2.2) with the following hypotheses:

H_1: The alleged father is the biological father.
H_2: The alleged father and the child are unrelated.

(a) Specify two allele systems, L1 and L2, with alleles 12, 13, and 14 for both systems.

Hint: `File->Frequency database`. Let $p_{12} = 0.1, p_{13} = 0.2$, and $p_{14} = 0.7$ for both systems. The loci are closely linked. Specify the recombination rate (θ) as 0.01. *Hint*: `File->Frequency database->Options`.

(b) In several situations, we have only the genetic distance between markers, measured in centimorgans (cM). How can we convert genetic distance to recombination rate?

(c) Select the appropriate pedigrees with `File->New wizard` and specify the data for the father as homozygous 12/12 for both loci and for the child as heterozygous 12/13 for both loci. Calculate the LR, which should coincide with the theoretical value of 25. Also compute the posterior probability with equal priors. *Hint*: Press the LR/Posterior button to change the displayed results.

(d) Try changing the recombination rate to 0.5 and calculate the LR again. What happens? Explain why.

(e) *By hand, show that LR = 25. Use the same notation as in Equation 4.3.

(f) The alleged father states that it could be his brother who is the true father of the child. Change H_2 to uncle by returning to the `Select alternative hypothesis` window. Calculate the LR with the recombination rate specified as 0.01.

(g) Save the `FamLink` file (we suggest the file extension sav). Exit `FamLink`.

Exercise 4.2 A case of disputed sibship. The exercise involves two persons P1 and P2 interested in finding out whether they share the same mother and/or father. We consider multiple hypotheses:

H_1: P1 and P2 are full siblings.
H_2: P1 and P2 are half siblings.
H_3: P1 and P2 are unrelated.

(a) Reuse the frequency database specified in the previous exercise. Specify the recombination rate to 0.09. *Hint*: `File->Open->Project`. Select the appropriate pedigrees, corresponding to the hypotheses above, and specify the data for P1 as homozygous 12/12 for both loci and P2 as homozygous 12/12 for both loci. *Hint*: Select multiple alternative hypotheses by holding `Ctrl` while making the selection.

(b) Calculate the LR. Scale versus the unrelated pedigree and discuss the LRs for full siblings and half siblings. Compute also the posterior probabilities with flat priors.

(c) *Compute by hand the LR comparing H_1 versus H_2.

(d) Try changing the recombination rate to 0.5 and calculate the LR again. What happens? Explain why.

(e) *Can linked autosomal markers be used to distinguish maternal from paternal half siblings?

Exercise 4.3 Immigration case. In cases of immigration, several alternative hypotheses may be relevant, and linked markers may prove useful in the determination of the most probable one. In the current exercise, we will explore a case where

Table 4.13 Allele Frequencies for Exercise 4.3

Alleles	vWA (Frequency)	D12S391 (Frequency)
14	0.30	
15	0.20	
16	0.05	
17	0.20	0.20
18	0.25	0.10
19		0.30
20		0.20
21		0.20

autosomal markers are unable to distinguish between the alternatives. Consider marker data for two markers vWA and D12S391, with frequency data as indicated in Table 4.13. We furthermore specify the following hypotheses:

H_1: P1 is the uncle of P2.
H_2: P1 is the grandfather of P2.
H_3: P1 and P2 are unrelated.

(a) Specify the recombination rate as 0.1. Specify the genotypes for P1 as 15/15 and 20/20 for vWA and D12S391, respectively, while P2 has identical genotypes as P1, for both loci. Calculate the LR and scale versus unrelated. Discuss the results.
(b) Discuss what would happen if you were to change the recombination rate to 0.5.
(c) Change the recombination rate to 0.5 and compare the result with your result from (b).
(d) Calculate the LR also for $\rho = 0.25$ and compare the results from (a), (b), and (c).
(e) Save the project with the .sav file extension.

Exercise 4.4 It was my brother who did it! `FamLink` may be used in criminal cases where the defendant claims: "It was my brother who did it!" Suppose we have some stain, assumed to be from one person, from a crime scene where we wish to provide some probability as to whether a suspect could be the source of the stain. The random match probability is low, suggesting that the defendant may be the source of the stain. The defendant claims he is innocent, and suggests that a brother of his committed the crime. We consider the following hypotheses:

H_{D1}: The brother of the defendant is the source of the profile in the stain.
H_{D2}: A random man, is the source of the profile in the stain.
H_P: The defendant is the source of the profile in the stain.

(a) Open the saved project from Exercise 4.3, containing the frequency database. Specify the recombination rate as 0.09.

(b) Select the pedigree `Full siblings` (corresponding to H_{D1}) as the main hypothesis and select `Unrelated` as the alternative hypothesis (corresponding to H_{D2})

(c) For the two individuals enter the same data for both individuals, 14/14 for vWA and 21/21 for D12S931.

(d) Calculate the LR. Discuss the results.

(e) *Compute the likelihood for H_P and H_{D2} by hand. Calculate also the LR comparing H_P versus H_{D2}.

(f) *Calculate the remaining LRs comparing H_P with H_{D1}. *Hint*: Use the LR calculated by hand and the one obtained in (e).

(g) Export the multiplication factor and discuss its meaning. *Hint*: Use `Save results`.

(h) Given the results in (e), the defendant is interested in knowing if some other relative could be blamed. As a scientist you are curious to find this out and try "Uncle" and "Grandfather" as alternative hypotheses. Calculate the LR and discuss the results.

Exercise 4.5 On the impact of linkage. The purpose of this exercise is to illustrate how `FamLink` can be used to study the general impact of accounting/not accounting for linkage in an incest case; see Exercise 2.6. To study the impact on a specific case we may use simulations, a resourceful tool included in the software. Simulations are performed by our assuming each of the specified hypotheses to be true, and we calculate the LR for each case. A summary statistic is then reported.

We specify the hypotheses as follows:

H_1: The father of the mother is also the alleged father of the child.

H_2: Another man, unrelated to the alleged father, is the father of the child. The alleged father is still the father of the mother.

(a) We use real frequency data from the STR markers vWA and D12S391. Import the file **Exercise4_5.txt**, containing the frequency database for the two STR markers. *Hint*: `File->Frequency database` and `Import`.

(b) Set the recombination rate to 0.089. (This corresponds to estimates in the literature). Select appropriate pedigrees but do not enter DNA data.

(c) Perform simulations with a seed value of 12345. *Hint*: Instead of `Calculate`, hit `Simulate`. Do 1000 simulations, which is generally a reasonable number to obtain an idea of the distribution of the LR. Select `Save raw data` and place the file somewhere you can find it.

(d) Once we have performed simulations, we may store a summary. Save a simulation report and explore the contents: `Save results` and `Simulation report`.

(e) What is the median effect of linkage on the LR in the current case given H_1 is true?

(f) *Open the raw data txt file in Excel (or similar software), and estimate the average, given H_1 is true, from the ratio

$$\text{LR(not accounting for linkage)/LR(accounting for linkage)}.$$

(g) *The LR if H_2 is true will often be zero in the current case. Explain why. *Note*: Zeros will be disregarded when calculating summary statistics.

(h) **The expected LR under H_2 is 1 as shown in Equation 8.3. Is this true in the current case? What can be done to improve the fit to the expectation?

Exercise 4.6 A real example. Another pair of closely located markers is SE33 and D6S1043 with a genetic distance of only 4.4 cM. The markers are both included in some commercial STR multiplexes. We are interested in finding the multiplication factor for the following hypotheses:

H_1: Two females, Sister1 and Sister2, are full siblings. (Data are available for the common mother.)
H_2: Maternal half siblings. (Data are available for the common mother.)

(a) Compute the recombination rate in FamLink with Haldane's mapping function. *Hint*: Tools->Conversion.

(b) We use real allele frequency data from a Chinese Han population. Import the file **Exercise4_6.txt**, containing the frequency database for the two STR markers. Specify the recombination rate calculated in (a).

(c) Specify the data for Sister1 as 14/14 for D6S1043 and 21/23.2 for SE33, and for Sister2 as 14/14 for D6S1043 and 21/24 for SE33. Furthermore, specify the data for the mother as 14/19.3 and 17/21.

(d) Calculate the LR in FamLink and compute the multiplication factor by hand from the results.

(e) The multiplication factor may be combined with the total LR obtained in some other software, where linkage is not accounted for. Discuss how this is possible.

(f) *Perform 1000 simulations (use seed=12345). Find the median for the multiplication factor in the simulation report for each simulated hypothesis. What are the values for the 5th and 95th percentiles? *Hint*: Use the fact that the multiplication factor is equal to LR(no linkage)/LR(linkage).

Exercise 4.7 Defining pedigrees. FamLink includes a number of predefined pedigrees where the user needs only to select the required pictures indicating the family structure. We may also create our own pedigrees with the Merlin input file notation (see http://www.sph.umich.edu/csg/abecasis/merlin/tour/input_files.html). Consider a case of three persons interested in knowing whether they are all full siblings or unrelated:

H_1: Three persons (P1, P2, and P3) are full siblings.
H_2: P1, P2, and P3 are unrelated.

(a) We will use the same frequency database as in Exercise 4.1. Specify the recombination rate as 0.1.

(b) Specify the needed relationships in a text file and name it **Exercise4_7.ped**. Specify the data for all persons as 12, 12 at the first locus in the same file. For the second locus we will add a previously unseen allele, denoted 11; specify the data for all persons as 11, 12.

(c) Import the pedigree into `FamLink`. (This is done by `File>New wizard>Import ped file`.) When the new allele is imported, we require a frequency. Specify 0.05 for allele 11 and select `Search and subtract` as the method. (See the manual at http://famlink.se for details on the "search and subtract" method.)

(d) **Discuss the search and subtract method in the current context. Would it be applicable in, for example, `Familias`?

(e) Calculate the LR and discuss the results.

(f) Discuss the hypotheses and whether you would consider alternatives.

Exercise 4.8 A case of identification. The following exercise will serve to illustrate the importance of linkage in a real case. Identification of unidentified remains is a recurring task in forensic laboratories. A skeleton was found and successfully genotyped with two overlapping commercial STR kits, yielding in total data for 23 autosomal STR markers. A putative sister of a missing person was genotyped for the same 23 markers. The hypotheses are defined as follows:

H_1: The sister is the biological sister of the unidentified remains.
H_2: The sister is unrelated to the unidentified remains.

The combined results indicated LR $= 500$ in favor of H_1. We will use `FamLink` to examine the results for the markers vWA and D12S391 and how the linkage between the two markers influences the results.

(a) Open the file **Exercise4_8.sav** in `FamLink`. The file contains a real frequency database for the indicated STR markers from a Norwegian population sample. Specify the recombination rate as 0.089.

(b) Select the hypotheses indicated above and enter the data for the remains as 14/14 for vWA and 21/21 for D12S391. Similarly for the sister enter 14/14 for vWA and 21/21 for D12S391.

(c) Calculate the LR and discuss the results. Would accounting for linkage change your conclusion?

(d) Try changing the genotype data for the persons to see how this affects the outcome.

(e) *In light of the results, discuss when we can expect an increase/decrease in the LR when accounting for linkage in the current case.

Exercise 4.9 *Theoretical exercise. The following exercise is meant for mathematically oriented readers. We will consider the well-discussed example of an uncle versus half siblings, mentioned earlier in the text. Consider two STR markers L1 and

L2, where the genetic distance and hence the recombination is unspecified but can be denoted by ρ. We further consider the following hypotheses:

H_1: P1 and P2 are related as half siblings.
H_2: P1 and P2 are related as uncle and nephew.

(a) *The genotype data are given in Table 4.14. Use your knowledge of IBD patterns to derive a theoretical formula for the LR for the two markers L1 and L2. *Hint*: Use methods described in Example 4.2.
(b) *Confirm your formula with FamLink, and use $p_{12} = 0.1, p_{21} = 0.05$, and $r = 0.1$. *Hint*: The frequencies of the other alleles are irrelevant; use any value.
(c) *Plot the LR as a function of the recombination rate with the same allele frequencies as in (b).

Table 4.14 Genotype Data for Exercise 4.10

	P1	P2
L1	9/12	12/15
L2	19/21	21/25

Exercise 4.10 Extended example. In addition to creating our own pedigrees, we may also analyze previous Familias projects (version 1.81 or above) to obtain an LR where linkage between virtually any number of markers is considered. All commonly used STR markers as well as a number of less common markers are predefined with their genetic distances specified; see the file **0markerInfo.ini** provided in the FamLink install directory. As indicated, we may now consider more markers and more complicated pedigree structures. The following is extracted from a real case (see Figure 4.18). Consider the following:

H_1: The alleged father is the biological father of the child.
H_2: The alternative father is the biological father of the child.

(a) Open the file **Exercise4_10.fam** in Familias and explore the project. (*Note*: The person marked as "cousin" in Figure 4.18 is not included since Familias versions below 3.0 cannot handle the complexity of the project.) The project contains genetic marker data for 35 autosomal STR markers.
(b) Calculate the LR in Familias. Be patient, the computations may take more than 20 min depending on the performance of your computer. You may also choose to skip this.
(c) Open the Quick analysis interface in FamLink. Select Generate report and browse for the file **Exercise4_10.fam**.
Hint: Tools->Quick analysis.
(d) Perform an analysis of the Familias project by pressing Analyze.

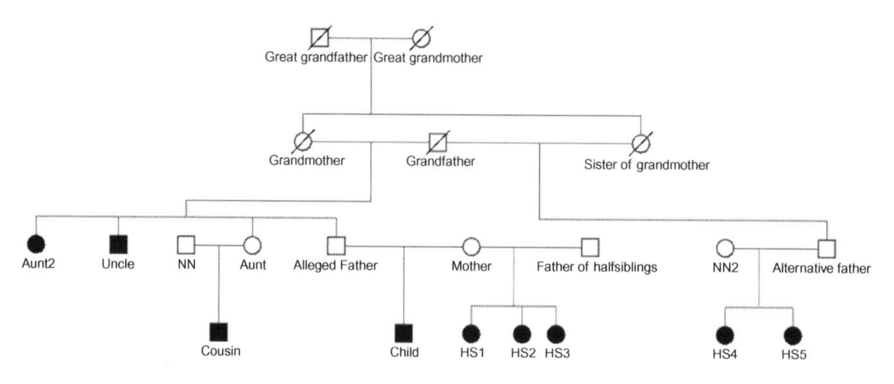

FIGURE 4.18

Complex kinship case.

(e) Browse the contents of the report file. Find the LR and compare it with the one obtained in `Familias`.

(f) The analysis includes markers from the commercial kits HDplex (QIAGEN), PP16 (Promega), and ESX17 (Promega). What can be said about the number of linked markers on different chromosomes obtained by use of all the kits? *Hint*: See `FamLink` report.

Exercise 4.11 **Combining linkage and subpopulation effects. This exercise is interesting for the more mathematically oriented users. To allow for subpopulation structures, we may adjust the allele frequencies by θ correction; see Section 2.5. We may actually combine the effect of linkage with correction for subpopulation structure in a model. Consider the following hypotheses:

H_1: Two persons, P1 and P2, are related as half siblings.
H_2: P1 and P2 are unrelated.

P1 and P2 have genotype data as indicated in Table 4.14.

(a) *Use the sampling formula (see Section 2.5.1), and derive the theoretical expression for the LR. *Hint*: You may reuse formulas derived in Exercise 4.9.

(b) *Use $p_{12} = 0.2, p_{21} = 0.1$, and $\rho = 0.01$, and compute the LR.

(c) **Use the sampling formula provided in Section 2.5.1 and $\theta = 0.02$, and compute the updated set of allele frequencies. *Hint*: Given zero alleles are IBD, we have four different observations, whereas given one allele is IBD, we have three different observations.

(d) **Compute the LR with $\theta = 0.02$. *Hint*: Use the results in (b) and (c).

(e) **Plot the LR versus θ with the specifications in (b).

4.5.2 X-CHROMOSOMAL MARKERS AND `FamLinkX`

`FamLinkX` implements an algorithm for linked markers on the X chromosome. In addition to linkage, the software accounts for LD (allelic association) and mutations. The software is intended to be user-friendly, but may provide obstacles for the inexperienced user. `FamLinkX` provides the LRs using three different methods: M1—exact model, considering linkage, LD, and mutations; M2—cluster approach (see the manual for Merlin), considering linkage and LD but not recombinations within clusters and not mutations; M3—only linkage is considered between markers. In the following exercises, we are interested in M1 as this is the preferred model, especially for STR markers, but comparisons with the other models will be made. For all calculations, unless stated otherwise, we consider X-chromosomal marker data and the corresponding inheritance patterns.

Exercise 4.12 A paternity case revisited. We will first revisit the paternity case (duo) with the following hypotheses:

H_1: The alleged father is the biological father of the female child.
H_2: The alleged father is a random man.

Open `FamLinkX` and specify the frequency database. *Hint*: `File->Frequency database`. Create a new cluster and specify two diallelic systems, L1 and L2, with alleles 12, 13 and 16, 17, respectively. Let $p_{12} = 0.6, p_{13} = 0.4$ for L1 and $p_{16} = 0.6$, $p_{17} = 0.4$ for L2. Select the `Simple mutation model` with the mutation rate set to 0 for both systems. Set the genetic position to 10 cM for L1 and 10.1 cM for L2. Furthermore, specify haplotype observations according to Table 4.15.

(a) Why is it important that we explicitly specify the gender of all persons in calculations for X-chromosomal marker data?
(b) What is the estimated recombination rate between the two loci? *Hint*: Use Haldane's mapping function.
(c) Why do we specify the number of observations for each haplotype?
(d) Use the equation below to calculate a measure of the association between the alleles:

$$r^2 = \frac{(p_{12}p_{16} - p_{12,16})^2}{p_{12}p_{13}p_{16}p_{17}}. \tag{4.12}$$

Table 4.15 Haplotype Observations for Exercise 4.12

	12	13
16	59	1
17	1	39

(e) Set λ to 0.0001 in `Options`. We will discuss the importance of λ in Exercise 4.14 and will not dwell further on it now. In brief, setting a low λ means we give much weight to the observed haplotypes in Table 4.15.

(f) Select the appropriate pedigrees with the wizard. The alternative hypothesis depicts two unrelated girls, but the first hypothesis will override the genders. *Hint*: `File->New wizard` and specify the data for the father as 12 for L1 and 17 for L2 and for the child as 12/12 for L1 and 17/17 for L2.

(g) Calculate the LR, which should coincide with the theoretical value of 100. Choose to save the file when asked. There may be a small deviation from the theoretical value, which we will return to in later exercises.

(h) Change the genotypes for the child to 12/13 for L1 and 16/17 for L2. Calculate the LR.

(i) Change the number of observations for each haplotype. What happens? Explain why.

(j) *Discuss the high degree of LD and if this is situation is likely to occur in reality.

Exercise 4.13 A case of sibship revisited. In this exercise, we revisit the example in Exercise 4.2 concerning disputed sibship. Two females, F1 and F2, are interested in finding out whether they are siblings in some way. We specify the following hypotheses:

H_1: F1 and F2 are full siblings.
H_2: F1 and F2 are maternal half siblings.
H_3: F1 and F2 are paternal half siblings.
H_4: F1 and F2 are unrelated.

(a) Explain why we may distinguish H_2 from H_3 with X-chromosomal markers but not with autosomal markers.

(b) Use the same frequency data, and haplotype observations, as in Exercise 4.12; alternatively, open the file **Exercise4_13.sav**.

(c) Specify $\lambda = 0.0001$ and select the relevant hypotheses. (*Note*: If you also stored the case-related DNA data in the previous exercise, you may be asked if you wish to erase all DNA data. Answer "yes.")

(d) Enter data for both F1 and F2 as 12/12 for L1 and 17/17 for L2. Calculate the LR.

(e) Scale against H_4. Comment on the importance of accounting for LD and linkage in the current case.

Exercise 4.14 The importance of λ. This exercise is intended to provide some insight into how the haplotype frequencies are estimated and the importance of the parameter λ. Our model for haplotype frequency estimation is described by

$$F_i = \frac{c_i + p_i\lambda}{C + \lambda}, \tag{4.13}$$

where F_i is the haplotype frequency for haplotype i, c_i is the number of observations for the haplotype, p_i is the expected haplotype frequency (under the assumption of linkage equilibrium), C is the total number of observations for all haplotypes, and λ is a parameter giving weight to the expected haplotype frequencies. This model allows unobserved haplotypes to be accounted for, in contrast to models which estimate the haplotype frequency solely on the basis of the counts. The difficulty lies in the choice of a good λ. Our recommendation is to compute the LR for a number of different values and select the least extreme value—that is, the value closest to 1. In Section 7.3 we explore methods to estimate λ.

To start, we specify a case with an aunt of a female child:

H_1: The female is the aunt of the child.
H_2: The two females are unrelated.

(a) Again use the same frequency data as in Exercise 4.12. Select the relevant hypotheses. *Hint*: Select the *Aunt/Uncle* as the main hypothesis.
(b) Enter data for the child as 12/12 for L1 and 16/17 for L2 and for the aunt as 12/13 for L1 and 17/17 for L2.
(c) Calculate the LR for $\lambda = 0.0001, 0.01, 1, 100$, and 10,000.
(d) What happens for large values of λ?
(e) What happens for small values of λ? *Hint*: Use the equation for haplotype frequency estimation above.
(f) Change the data for the child to 13/13 for L1. Repeat (c) and discuss the results.

Exercise 4.15 Extended example. (Combining Familias and FamLinkX). This exercise provides a challenge where the user needs to combine the results from Familias and FamLinkX to obtain a final result. The data are extracted from a real case (anonymized) where three females provided DNA samples. The hypothesis is:

H_1: The three females are all full siblings, versus all other alternatives (H_2).

Obviously H_2 cannot be used in the current setting, in a simple way, and we need to refine possible alternative hypotheses. Tasks (a), (b), and (c) involve the use of Familias; however, you may also skip to task (d) for FamLinkX.

(a) *Open the file **Exercise4_15.fam** in Familias 3. We assume that all females are children with no children of their own. Specify that the three females are children. Also define two parents, a mother and a father, and specify that they were both born in 1970.
(b) *Use Familias to find the pedigrees with a posterior probability above 0.001. What are the most probable relationships according to the results? Interpret the results. *Hint:* Use the generate pedigrees function in Familias.
(c) *Discuss the constraints specified in (a) and their impact on the results in (b).
(d) Open the database **Exercise4_15.sav** in FamLinkX, which contains frequency and haplotype data for the Argus X12 kit from QIAGEN based on a Swedish population sample. Explore the haplotype frequency database.
(e) On the basis of the results in (b), we specify the following hypotheses:

H_1: The three females are all full siblings.
H_2: Two females are full siblings and the third female (named F3) is a paternal half sibling.
H_3: Two females are full siblings and the third female (named F3) is a maternal half sibling. Draw the pedigrees corresponding to the hypotheses specified above.

(f) Import the DNA data, available in **Exerecise4_15.txt**. Make sure to import the data in the file to the correct corresponding persons; the person denoted F3 should be imported to 3.
(g) Calculate the LR and interpret the results. Be patient, the computation may require more than 20 min. *Hint*: To speed up the computations, go into `File->Advanced`, and select and edit the hypothesis we have chosen to investigate. For each pedigree set the `Threshold` value to 0.001 and the `Steps` value to 0.
(h) Discuss if the LR in (g) may be combined with the results in (b)? What is your final conclusion on the case?

Exercise 4.16 Further discussion of λ. We will provide an example of how the value of λ may be crucial to the conclusion in a case. We use anonymized data from a real case with two typed females (F1 and F2) and consider the following hypotheses:

H_1: F1 and F2 are paternal half siblings, with different mothers.
H_2: The two females are unrelated.

(a) Open the file **Exercise4_16.sav**, containing the frequency database.
(b) Select appropriate pedigrees and import genotype data from the file **Exercise4_16.txt**.
(c) Compute the LR for a number of different values of λ—for example, 0.001, 1, 100, and 1000.
(d) Discuss the results in (c). What conclusion can be drawn?
(e) *Explore the genotype data and see if you can find an explanation for the results. Use your knowledge of haplotype phases under the different hypotheses. *Hint*: Use the frequency estimation tool in the `Edit cluster` dialog.
(f) Discuss what value of λ should be chosen. What is the most conservative value?

Exercise 4.17 English Speaking Working Group 2013 paper challenge. This exercise covers the calculation of the LR for the X-chromosomal data included in the English Speaking Working Group 2013 paper challenge [91]. The question involved a paternity duo with the following hypotheses:

H_1: The alleged father is the biological father of a female child.
H_2: The alleged father and the child are unrelated.

(a) Open the file **Exercise4_17.sav** containing the frequency database. Explore the specification of the haplotypes. Make sure λ is set to 1.
(b) Select appropriate pedigrees and import genotype data from the file **Exercise4_17.txt**.
(c) Calculate the LR.
(d) Change the value of λ to 212 and see how this affects the results.
(e) Compare the results calculated with the LE model (M1) with the results calculated with the exact model (M3).
(f) **Try deriving the algebraic formula for the two markers in the first cluster. For simplicity, assume mutations can be disregarded.

Exercise 4.18 *Creating pedigrees. `FamLinkX` includes some predefined pedigrees where the calculations are performed with method M3. In addition, you may wish to create new pedigrees. The implementation currently does not allow method M3 to be used on user-defined pedigrees; therefore, calculation with only methods M1 and M2 will be done. Two females F1 and F2 are related as maternal cousins. In addition, we are asked to find out whether they also share the same father:

H_1: F1 and F2 are maternal cousins as well as paternal half siblings.
H_2: F1 and F2 are maternal cousins.

(a) Open the file **Exercise4_18.sav** containing the frequency database. We will consider different values of λ. Start by setting λ to 1.
(b) Continue by creating the pedigree for H_1. You need to define several extra persons to specify the necessary relations.
Hint: `Tools->Select pedigree->Create/Edit pedigree`.
  1. Rename the pedigree as H1.
  2. Add extra persons *Hint*: Use the `Persons` button within the edit pedigree dialog. Add persons named F1, F2, grandmother, grandfather, mother1, mother2, and father. Make sure the genders are correct for all individuals.
  3. Specify the relations in H1.
  4. Close the edit pedigree dialog. The created pedigree will appear last in the list of pedigrees. Select the created pedigree and `Next`.
  5. Create the alternative hypothesis, H_2, with use of the same procedure as for H1. *Note*: You need not define any new persons; use the same as used to create H1.

(c) Import genotype observations from the file **Exercise4_18.txt**. Make sure you import the person named *cousin1* to F1 and the person named *cousin2* to F2.
(d) Calculate the LR.
(e) Repeat the calculations for different values of λ. What is your conclusion regarding the relationship between the two women.

Exercise 4.19 *Theoretical considerations. For verification purposes, and to validate the implementations, it may be interesting to derive theoretical formulas. Generally this is infeasible, but for some cases, with use of simplifications, algebraic formulas can be derived. We will start by considering one diallelic SNP marker, S1, with alleles 1 and 2. The frequencies are $p_1 = 0.4$ and $p_2 = 0.6$ and the mutation rates are zero. Furthermore, we specify the following:

H_1: A woman is the alleged mother of a child.
H_2: The alleged mother and the child (girl) are unrelated.

(a) The alleged mother has genotype 1/1, while the child has genotype 1/2. Compute the LR by hand and verify the answer in `FamLinkX`.
The alleged mother is discarded as the true mother for reasons other than the DNA. Instead an alleged father is presented. The hypotheses become the following:

H_1: A man is the alleged father of a girl.
H_2: The alleged father and the girl are unrelated.

(b) The alleged father has genotype 1, while the child still has genotype 1/2. Compute the LR by hand and verify it in `FamLinkX`. (*Note*: The alternative hypothesis displays two unrelated females, while as pointed out earlier, the genders in the main hypothesis override this information.)
(c) A second SNP, S2, is introduced with alleles 3 and 4. The frequencies are $p_3 = 0.6$ and $p_4 = 0.4$. The alleged father has genotype 3, while the child has genotype 3/3. Compute the LR by hand and verify it in `FamLinkX`. You may assume that the two SNPs are located far from each other and thus the calculations are independent. This is in principle impossible in `FamLinkX`. However, a sufficient approximation is to define the genetic position well apart—for example, 500 cM.
(d) S1 and S2 are in fact closely located, separated by only 0.1 cM. Specify this in `FamLinkX`. Calculate the LR again.
(e) There are further haplotype observations according to Table 4.16. Specify these in `FamLinkX`.
(f) Use three different values for λ (0.01, 1, and 100), and compute the LR in `FamLinkX` for each of them.
(g) *Derive the theoretical formula to confirm the calculations in (f) for $\lambda = 100$.
(h) **Plot the LR as a function of λ. *Hint*: Use the formula in Exercise 4.14.
(i) **Try to derive the formula for the paternity case in Exercise 4.12 (g) with the specifications in (e) and (f) in this exercise.

Table 4.16 Haplotype Observations for Exercise 4.19

	1	2
3	1	59
4	39	1

Exercise 4.20 **Exploring the algorithm. The exercise will explore the algorithm implemented in `FamLinkX`. Kling et al. [6] provide further details that may be needed to solve some of the questions. We first introduce one X-chromosomal marker with alleles 12, 13, 14, 15, and 16. Consider the following hypotheses:

H_1: Two girls, F1 and F2, are maternal half siblings.
H_2: F1 and F2 are unrelated.

(a) *Given H_1, specify the founder allele patterns—that is, a possible set of alleles for the founders—if the genotypes are 12/13 for F1 and 12/15 for F2. Assume the mutation rates are zero. *Hint*: Plot the pedigree on paper and specify the founders.
(b) *Given H_1, specify the founder allele patterns if the genotypes are 12/13 for F1 and 14/15 for F2. Now assume one-step mutations are possible.
(c) *Set up the inheritance patterns—that is, which meioses must be accounted for?
(d) *Use the results in (a) and (c), and compute the number of different combinations we have to consider (in other words, the number of founder allele patterns times the number of inheritance patterns).
(e) **Now consider a second marker with the same data—that is, the results in (d) apply also for this marker. The two markers are linked such that for each combination for the second marker we also have to consider the different inheritance patterns in (c). With this information, compute the updated number of combinations we have to consider for the second marker.
(f) **We also need to account for LD with use of founder allele patterns—that is, for each combination at the second marker we have to consider all possible states at the first marker. With this information, compute the updated number of combinations we have to consider for the second marker.
(g) **Finally, we consider a third marker, tightly linked with the second marker. Given that we have to consider LD across all three markers, how many combinations do we have to consider for the third marker? *Hint*: Assume that linkage stretches only one marker—that is, given a specific inheritance state for the second marker, the third marker is independent of the different marker states at the first marker.
(h) **Consider instead H_2 and compute the number of combinations for the third marker. What effect does linkage have on the number of combinations we have to consider?

Exercise 4.21 A case with mutation. We will consider a case where we introduce a possible mutation. Consider two females, F1 and F2, with the following hypotheses about the relationship:

H_1: Maternal half siblings.
H_2: Paternal half siblings.

These are not distinguishable with autosomal markers.

(a) Open the file **Exercise4_21.sav**.

(b) Select maternal half siblings as the main hypothesis and select paternal half siblings as the alternative.
(c) Import genotype data from the file **Exercise4_21.txt**.
(d) Compute the LR. Use the default $\lambda = 1$.
(e) Change the data for the first individual (F1) to 19/25.1 for the marker DXS10148.
(f) Compute the LR. Use default values for mutation parameters—that is, the simple model where each mutation is equally likely.
(g) Change the data for the first individual (F1) to 19/19 and compute the LR.
(h) Change the mutation model to `Extended model` with the `Range` parameter equal to 0.1 and the `Rate 2` parameter equal to 0.000001.
(i) Compute the LR for the data as indicated in (e) and (g).
(j) *Discuss the results and the importance of good models for mutations. Why is this decisive in the current case?

Exercise 4.22 *Understanding complex parameters. We will now look more closely at some parameters that may further affect the results in `FamLinkX`, *Threshold* and *Step*. The former, denoted here by t, is between 0 and 1, and specifies the minimum value required for the overall pedigree likelihood to be included in the calculations. In other words, this limit applies to $\Pr(D_i \mid V_i, F_i)$ in Equation 4.9.

The latter parameter, denoted here by s, is a non-negative integer, and specifies the number of steps we consider for founder alleles. For instance, in a case of paternal half sisters, we need to sum over the possible alleles for the common father, which is given by the shared alleles of the sisters. If we consider mutations, all alleles are possible, while we can use the *Step* parameter to define the number of steps away from the sisters' alleles that we wish to include in the summation. In reference to Equation 4.9, this limits the founder allele space F_i.

We will use an example to illustrate the effect of tuning the different parameters. Consider the following:

H_1: Two females, F1 and F2, are paternal half sisters.
H_2: The two females are unrelated.

(a) Open the file **Exercise4_22.sav**. The file contains a database for a Somali population [7]. The extended stepwise mutation model is specified for all markers.
(b) Specify the hypotheses according to H_1 and H_2. Import case data from the file **Exercise4_22.txt**.
(c) Calculate the LR. Make sure λ is set to 1.
(d) Explore the genotype data to find an explanation for the results.
(e) In the advanced settings (`File->Advanced`), change the *Threshold* parameter for H_1 to 1-e6, leaving the *Step* parameter unchanged for both H_1 and H_2. Save the data.
(f) Compute the LR.
(g) Change the *Threshold* parameter for H_1 to 0.0001. Compute the LR.

(h) *What can be said about the effect of the *Threshold* parameter?

(i) Change the *Threshold* parameter to 0.0001 and the *Step* parameter to 1 for H_1. Compute the LR.

(j) Change the *Step* parameter to 2 for H_1 and compute the LR.

(k) Change the genotypes for F1 to 25/30 for the locus DXS10101. Compute the LR for *Step* =0,1 and 2 (use *Threshold* =1e-8). Comment on the results.

(l) *What can be said about the effect of the *Step* parameter?
We recommend keeping the value of *Threshold* high (e.g., 0.001-0.00001) and the value of *Step* low (e.g., 0 or 1) as this reduces the computational effort considerably compared with when we have lower values for the *Threshold* parameter and higher values for the *Step* parameter. If the LR is still zero for all computation models, a lower value on the *Threshold* parameter may be considered.

CHAPTER

Relationship inference with R 5

CHAPTER OUTLINE

R[1] is a powerful tool for programming, plotting, and statistical analysis. The core source code of Familias is available[2] through R and access functions are presented in this chapter. The R version allows users to extend the functionality described in previous chapters and make their contribution available to the community. We also present other useful R packages.

There is a large and growing literature on R and its applications to statistics. A list of books is available at http://www.r-project.org/doc/bib/R-books.html. Currently, there is no book specifically dedicated to forensics genetics, but [92] addresses forensic applications more generally.

5.1 USING R

In previous chapters, computations were done either by hand or with use of various Windows programs: for example, Familias, Famlink, and FamlinkX. As with many Windows programs, these programs have user interfaces that focus on helping the user to apply the programs correctly. They are also meant to be easy to use for beginners. However, the very properties that support such features can also make the programs less flexible. Researchers who want to pose new questions or update or change what is computed will often prefer tools where they can combine existing methods and computer code with their own questions and methods. In the area of statistical computations, R is one of the most widely used such tools. It is free and open source, and can be downloaded from http://www.r-project.org. It does require

[1] http://www.r-project.org/.
[2] http://cran.r-project.org/web/packages/Familias/.

some work to get started with it, but once the basics have been mastered, users generally find it extremely powerful and convenient. An important feature of R is that it is fairly easy for users to contribute "packages" of code and data to databases from which other users can download and use them. The authors of this book have contributed the Familias package, which implements many of the same core functions as the Windows version, but also some other functions. A number of other packages with useful functions for pedigree inference exist[3] on the Comprehensive R Archive Network. Packages including DNAtools, DNAprofiles, kinship2, and paramlink are presented in Section 5.1.1 and in exercises in this and subsequent chapters.

In the remaining chapters of this book, we will use R for most computational examples. We will show how many of the computations discussed in previous chapters can also be done with the help of the packages mentioned above. But we also show how we can deal with additional questions and issues using R. Several issues are reexamined and presented from a more theoretical and general viewpoint.

Although the remaining chapters of this book can be read without knowing R, we encourage all readers not yet familiar with the program to download it and start using it. A number of tutorials can be found online; a very good possibility is "An Introduction to R," which can be found via "Help" and "Manuals" once R has been downloaded. Have a look at "Appendix A, A sample session" in that document to get a feel for what is possible with R. In what follows, we will assume you know the basics.

5.1.1 R PACKAGES FOR RELATIONSHIP INFERENCE

There are several packages available to solve the problems addressed in this book. Most of these packages have not been developed specifically for relationship inference. Their development has rather been motivated by applications in human genetics (linkage), population genetics and other forensic applications (mixtures, database problems). Some R packages available from http://www.r-project.org are listed below along with corresponding exercises. The exercises indicate a very small part of the functionality provided.

disclap,disclapmix A Y chromosome example is presented in Exercise 5.7.

DNAprofiles Simulation and importance sampling are exemplified in Exercises 8.7, 8.8, and 8.9.

DNAtools The package is developed to analyze forensic DNA databases as exemplified to a very small degree in Exercise 5.8.

Familias Section 5.1.2 presents this package. The exercises include Exercises 5.1, 5.2, 5.3, and 7.1.

identity Identical-by-descent probabilities and Jacquard coefficients of identity are calculated as demonstrated in Exercise 6.5.

[3] As of spring 2015.

`kinship2` This package is not used directly, but `Familias` and `paramlink` rely on the plotting functions provided.

`paramlink` As the name indicates, this package was originally intended for parametric linkage analysis. However, general functionality for plotting, likelihood calculation, and simulation are exemplified in Exercises 5.4, 5.5, 5.6, 6.4, 8.3, 8.5, and 8.6.

`BookEKM` is available from http://familias.name/book.html. It contains functions and code for examples in this book. Exercise 6.6 uses a function which is essentially a wrapper of code in `paramlink`.

5.1.2 THE FAMILIAS PACKAGE

The `Familias` R package contains in essence the same core code as the `Familias` Windows version, but instead of a Windows user interface, one can access the algorithm via R functions. This book uses version 2.3 of the package. If you have downloaded `Familias` from the Comprehensive R Archive Network and installed it within R, its use can be illustrated with the following simple paternity case computations (computer output is marked with a gray background):

```
library(Familias)

# Define the persons involved in a standard paternity case:
persons <- c("mother", "child", "AF")
sex     <- c("female", "female", "male")

# Define the two alternative pedigrees:
ped1 <- FamiliasPedigree(id = persons, dadid = c(NA, NA, NA),
       momid = c(NA, "mother", NA), sex=sex)
ped2 <- FamiliasPedigree(id = persons, dadid = c(NA, "AF",
      NA), momid = c(NA, "mother", NA), sex=sex)
plot(ped1); dev.new(); plot(ped2)
pedigrees <- list(notFather= ped1, isFather= ped2)

# Define a marker with four alleles with the given
# frequencies:
frequencies <- c(0.1, 0.2, 0.3, 0.4)
allelenames <- c("A", "B", "C", "D")
marker <- FamiliasLocus(frequencies, allelenames)

# Input the observed data
mother <- c("A", "B")
child  <- c("A", "C")
AF     <- c("C", "D")
datamatrix <- rbind(mother, child, AF)
#Compute the posteriors and likelihoods:
```

```
result <- FamiliasPosterior(pedigrees, marker, datamatrix)
result
```

```
$posterior
notFather  isFather
    0.375     0.625

$prior
notFather  isFather
      0.5       0.5

$LR
notFather  isFather
 1.000000  1.666667

$LRperMarker
      notFather isFather
locus         1 1.666667

$likelihoods
notFather  isFather
  0.00144   0.00240

$likelihoodsPerSystem
      notFather isFather
locus   0.00144   0.0024
```

As we can see, pedigrees may be constructed with the function FamiliasPedigree, where you need to submit a list of names of persons and their gender, together with the father and mother of each person, if in the pedigree. In our case, there are three persons involved, the mother, the child, and the alleged father (AF), and we compare the pedigree where the AF is the father with the one where he is not. Note how the plot function can be used to check that the pedigrees have been correctly constructed.

A locus is generated with the function FamiliasLocus. In its simplest form, it takes as input only a list of frequencies for the alleles, but additional properties about the probabilities of various types of mutations may also be specified. In our toy example, the genotypes are A/B, A/C, and C/D for the mother, child, and AF, respectively. The result object result is computed with the FamiliasPosterior function. Displaying this object shows that it has six components: FamiliasPosterior shows the posterior probability of each pedigree, given prior equal prior probabilities (the prior probabilities may optionally be specified differently). For reference, the prior probabilities used are also listed. The likelihood ratios (LRs) computed are listed in the LR component; by default, the LRs are computed comparing all pedigrees with the first. If there is more than one marker, the `LRperMarker` component may be useful. Similarly, the `likelihoods` and `likelihoodsPerSystem` list likelihoods. As an exercise, you should verify by manual calculations that the computed results are correct.

To illustrate a few more possibilities with the basic functions in the `Familias` package, we include a second, slightly more complex example, recomputing the results summarized in Table 2.5. First mutations are ignored.

```
persons <- c("child", "AF")
sex     <- c("male", "male")

ped1 <- FamiliasPedigree(id = persons, dadid = c(NA, NA),
      momid = c(NA, NA), sex=sex)
ped2 <- FamiliasPedigree(id = persons, dadid = c("AF", NA),
     momid = c(NA, NA), sex=sex)
pedigrees <- list(notFather= ped1, isFather= ped2)

data(NorwegianFrequencies)
# We will use the three loci D3S1358, D6S474, and TPOX.
D3S1358 <- FamiliasLocus(NorwegianFrequencies$D3S1358,
                         name = "D3S1358")
D6S474 <- FamiliasLocus(NorwegianFrequencies$D6S474,
                         name = "D6S474")
TPOX    <- FamiliasLocus(NorwegianFrequencies$TPOX,
                         name = "TPOX")
markers <- list(D3S1358, D6S474, TPOX)
# List the observed data, in pairs ordered according to the
# list above.
child <- c(17, 17,  16, 17, 8, 8)
AF    <- c(17, 18, 14, 15, 8, 8)
datamatrix <- rbind(child, AF)

#Compute the posteriors and likelihoods:
result <- FamiliasPosterior(pedigrees, markers, datamatrix)
result
```

```
$posterior
notFather  isFather
        1         0

$prior
notFather  isFather
      0.5       0.5
```

```
$LR
notFather  isFather
        1         0

$LRperMarker
        notFather isFather
D3S1358         1 2.450403
D6S474          1 0.000000
TPOX            1 1.805354

$likelihoods
   notFather     isFather
8.296258e-07 0.000000e+00

$likelihoodsPerSystem
          notFather    isFather
D3S1358 0.002368082 0.005802755
D6S474  0.003721643 0.000000000
TPOX    0.094134933 0.169946848
```

The LR in favor of fatherhood is zero, as there is an exclusion in the D6S474 marker. However, we can include a mutation rate of 0.001 and a default "Stepwise" model with rate of 0.5 as follows:

```
D3S1358_mut <- FamiliasLocus(NorwegianFrequencies$D3S1358,
                   maleMutationRate = 0.001)
D6S474_mut <- FamiliasLocus(NorwegianFrequencies$D6S474,
                  maleMutationRate = 0.001)
TPOX_mut   <- FamiliasLocus(NorwegianFrequencies$TPOX,
                  maleMutationRate = 0.001)
markers_mut <- list(D3S1358_mut, D6S474_mut, TPOX_mut)
```

Note how the `FamiliasLocus` function can be used both to create objects representing a locus and to edit male and female mutation parameters of such objects. Computations now give

```
result <- FamiliasPosterior(pedigrees, markers_mut,
      datamatrix)
result
```

```
$posterior
  notFather    isFather
0.995178312 0.004821688

$prior
notFather  isFather
      0.5       0.5
```

```
$LR
  notFather    isFather
1.000000000 0.004845049

$LRperMarker
                              notFather    isFather
NorwegianFrequencies$D3S1358          1 2.448882497
NorwegianFrequencies$D6S474           1 0.001096989
NorwegianFrequencies$TPOX             1 1.803548324

$likelihoods
   notFather     isFather
8.296258e-07 4.019578e-09

$likelihoodsPerSystem
                               notFather     isFather
NorwegianFrequencies$D3S1358 0.002368082 5.799154e-03
NorwegianFrequencies$D6S474  0.003721643 4.082603e-06
NorwegianFrequencies$TPOX    0.094134933 1.697769e-01
```

We see how the rightmost column of Table 2.5 is reproduced for these markers. As a final illustration, we recompute the results above with a nonzero theta correction (kinship parameter):

```
result <- FamiliasPosterior(pedigrees, markers_mut,
          datamatrix, kinship = 0.05)
result
```

```
$posterior
 notFather   isFather
0.99615292 0.00384708

$prior
notFather  isFather
      0.5       0.5

$LR
  notFather    isFather
1.000000000 0.003861937

$LRperMarker
                              notFather    isFather
```

```
NorwegianFrequencies$D3S1358                 1 1.870935641
NorwegianFrequencies$D6S474                  1 0.001270198
NorwegianFrequencies$TPOX                    1 1.625080437

$likelihoods
   notFather      isFather
1.083615e-06 4.184853e-09

$likelihoodsPerSystem
                                notFather     isFather
NorwegianFrequencies$D3S1358 0.003352028 6.271429e-03
NorwegianFrequencies$D6S474  0.002762636 3.509095e-06
NorwegianFrequencies$TPOX    0.117015584 1.901597e-01
```

For details about the current functionalities and options of the `Familias` package, consult the help function for the package, by writing

```
help(package = Familias)
```

5.2 EXERCISES

Exercise 5.1 Introduction to R `Familias`.

We begin with computations for a simple paternity case in order to illustrate the use of the `Familias` package, and compare it with computations by hand.

(a) Consider a paternity case where observations have been made at only one locus. The possible alleles at that locus are denoted A, B, C, and D, with population frequencies 0.1, 0.2, 0.3, and 0.4, respectively. The observed genotypes for the mother, child, and AF are A/B, A/C, and C/D, respectively. Disregarding any kind of complexities, such as population substructure, frequency uncertainties, and mutations, compute by hand the LR in favor of paternity.

(b) Use R to run the following code:

```
# Start the Familias package
# It must be already downloaded with e.g.
# install.packages("Familias")
library(Familias)

# Define the persons involved in a standard paternity
# case:
persons <- c("mother", "child", "AF")
sex     <- c("female", "female", "male")

# Define the two alternative pedigrees:
ped1 <- FamiliasPedigree(id = persons, dadid = c(NA, "AF",
```

```
    NA), momid = c(NA, "mother", NA), sex=sex)
ped2 <- FamiliasPedigree(id = persons, dadid = c(NA, NA,
    NA), momid = c(NA, "mother", NA), sex = sex)
pedigrees <- list(notFather = ped2, isFather = ped1)
# Note that the package by default will compute LR's using
# the first listed pedigree in the denominator, so
# notFather
# should be listed first to obtain the same result
# as in the manual calculation.

# Define a marker with four alleles with the given
# frequencies:
frequencies <- c(0.1, 0.2, 0.3, 0.4)
allelenames <- c("A", "B", "C", "D")
marker <- FamiliasLocus(frequencies, allelenames)

# List the observed data
mother     <- c("A", "B")
child      <- c("A", "C")
AF         <- c("C", "D")
datamatrix <- rbind(mother, child, AF)
# Note how the commands above create a matrix with the
# row names equal to the names of the persons involved.

#Compute the posteriors and likelihoods:
result <- FamiliasPosterior(pedigrees, marker, datamatrix)
```

Now, write `result` in `R`, and interpret the information in the result object created with the commands above. Use the command `help(FamiliasPosterior)` to help in interpreting the output. Check that you get the same result as you computed manually above.

(c) The remains of a man have been found, and it is speculated he is the missing uncle of two (full) brothers. More precisely, if we call the remains "body" and the two brothers B1 and B2, the hypothesis is that the father of B1 and B2, let us call him "father," is the full brother of "body," whereas the counterhypothesis is that "body" is completely unrelated to B1 and B2. Write `R` code as above to specify the two alternative pedigrees: In addition to "body," B1, B2, and "father," you should in both pedigrees include the persons "mother" (of B1 and B2), and "grandma" and "grandpa." *Hint*: Use `help(FamiliasPedigree)` to understand how to formulate the input for this function.

(d) Assume DNA data for the same locus as above have been obtained, and that one has observed A/A, A/B, and A/C for the body, B1, and B2, respectively. Use the `FamiliasPosterior` function to compute the LR in favor of the body being the missing uncle.

(e) In our simple context, where we assume that mutations are impossible, there are exactly five different genotypes that "father" could have. Find them. Then, for each of them, add it as observed data, and compute the likelihood of the two pedigrees with `FamiliasPosterior`. Show that the likelihood computed in the previous exercise equals the sum of the five likelihoods computed in this way.

(f) For the case that "father" has the genotype A/A and the hypothesis is that the body is the uncle, compute manually the likelihood for all the data, and compare it with the result obtained in the previous exercise.

Exercise 5.2 Mutation models. `R Familias`. Below you are asked to do LR calculations in `Familias` involving mutations and then verify the answer using a formula. Parts of Exercises 2.7 and 2.8 are similar for the Windows version of `Familias`. Consider the following hypotheses:

H_1: The AF is the father of a child.
H_2: The AF and the child are unrelated.

There is one marker with alleles and allele frequencies defined by

```
alleles <- 14:21
p <- c(0.072, 0.082, 0.212, 0.292, 0.222, 0.097, 0.020, 0.003)
```

In the below numerical examples, the AF is 14/15, while the child is 16/17. The mutation rate and the range are $R = 0.005$ and $r = 0.5$, and the same model is assumed for females and males.

(a) Calculate the LR for the "equal" mutation model—for instance by entering

```
library(Familias)
persons <- c("CH", "AF")
sex <- c("male", "male")
ped1 <- FamiliasPedigree(id = persons, dadid <- c("AF",
     NA), momid = c(NA, NA), sex = sex)
ped2 <- FamiliasPedigree(id = persons, dadid = c(NA, NA),
                               momid = c(NA, NA), sex = sex)
mypedigrees <- list(unrelated = ped2, isFather = ped1)
alleles <- 14:21
p <- c(0.072, 0.082, 0.212, 0.292, 0.222, 0.097, 0.020,
  0.003)
CH <- c(16, 17)
AF <- c(14, 15)
datamatrix <- rbind(CH, AF)
locus1 <- FamiliasLocus(p, alleles, "locus1",
  maleMutationModel = "equal", femaleMutationModel =
  "equal",
  maleMutationRate = 0.005, femaleMutationRate = 0.005)
FamiliasPosterior(mypedigrees, locus1 , datamatrix)
```

Calculate the LR for the other mutation models—that is, for "Proportional" and "Stepwise."

(b) Denote the genotypes of the AF and the child by a/b and c/d. We use a, b, c, and d to denote allele variables, so their values may or may not be equal. The LR may be written (e.g., Example 6.6)

$$\mathrm{LR} = \frac{1}{4}\frac{(m_{ac} + m_{bc})p_d + (m_{ad} + m_{bd})p_c}{p_c p_d}, \tag{5.1}$$

where m_{kl} is the probability that allele k ends up as allele l. Show that for the "equal" mutation model

$$\mathrm{LR} = \frac{1}{2}\frac{m(p_d + p_c)}{p_c p_d},$$

where $m = R/(n-1) = 0.005/7$. Use this to confirm the result for the "Equal" model in (a) above.

(c) The mutation matrix for the "Stepwise" model can be obtained as follows:

```
r <- 0.5
R <- 0.001
locus1 <- FamiliasLocus(p, alleles, "locus1",
            maleMutationModel = "Stepwise",
            femaleMutationModel = "Stepwise",
            femaleMutationRange = r, maleMutationRange = r,
            maleMutationRate = R, femaleMutationRate = R)
locus1$femaleMutationMatrix
```

Verify the result in (a) for the "Stepwise" model.

(d) General mutation models can be specified. Describe the model specified by the mutation matrix

```
M <- matrix(c(
0.9950,0.0050,0      ,0      ,0      ,0      ,0      ,0      ,
0.0025,0.9950,0.0025,0      ,0      ,0      ,0      ,0      ,
0      ,0.0025,0.9950,0.0025,0      ,0      ,0      ,0      ,
0      ,0      ,0.0025,0.9950,0.0025,0      ,0      ,0      ,
0      ,0      ,0      ,0.0025,0.9950,0.0025,0      ,0      ,
0      ,0      ,0      ,0      ,0.0025,0.9950,0.0025,0      ,
0      ,0      ,0      ,0      ,0      ,0.0025,0.9950,0.0025 ,
0      ,0      ,0      ,0      ,0      ,0      ,0.0050,0.9950),
8, 8, byrow = TRUE)
```

Calculate the LR for this model with `Familias` and confirm the result.

Exercise 5.3 Exercise 2.3 revisited. The Windows version (3.1.8 and above) of `Familias`, introduced in Chapter 2, implements a function to export a complete project to R Familias, thus simplifying analysis in the R version. The function mentioned is found in `File->Export to OpenFamilias`.

Redo Exercise 2.3 in R Familias with the functionality indicated above, and verify that the results are reproduced. (*Note*: There may some small rounding error.) *Hint*: Make sure `Plot pedigrees` is selected to create an illustration of the pedigrees. The latter functionality is useful to confirm and plot pedigrees defined in Windows Familias. Also, make sure that the `ref` parameter supplied to the `FamiliasPosterior` function is set to 2.

Exercise 5.4 Symmetric pairs, linked markers. `paramlink`. Some details from Section 6.4.1, in which the notation is described more fully, are addressed below. Assume genotype data are available for two persons—say, P_1 and P_2. We consider the following hypotheses:

H_1: P_1 and P_2 constitute a grandparent-grandchild relationship (as 1 and 6 in Figure 6.6).
H_2: P_1 and P_2 are half siblings (as 6 and 8).
H_3: P_1 and P_2 constitute a uncle-nephew relationship (as 3 and 6).

Thompson [93] gives the expressions

$$k_{1,1}^{(1)} = \frac{1}{2}(1-\rho),$$
$$k_{1,1}^{(2)} = \frac{1}{2}(\rho^2 + (1-\rho)^2) = \frac{1}{2}R, \text{ say,}$$
$$k_{1,1}^{(3)} = \frac{1}{2}((1-\rho)R + \rho/2),$$

for P_1 and P_2 to share one allele identical by descent at both markers as a function of the recombination rate.

(a) Plot the kappa functions.
(b) Use `paramlink` to plot the pedigrees corresponding to the hypotheses.
(c) Assuming the two individuals are homozygous for different alleles, show that

$$\text{LR}_{1,2} = \frac{L(H_1)}{L(H_2)} = \frac{1-\rho}{\rho^2 + (1-\rho)^2}.$$

Use `twoMarkerDistribution` of `paramlink` to check the result for a pair of SNP markers separated by $\rho = 0.29$.

Exercise 5.5 Plot for "A fictitious paternity case." `paramlink`. Use `paramlink` to make Figure 2.14. A suggestion for the commands follows:

```
require(paramlink)
H1 <- nuclearPed(2)
H1 <- addOffspring(H1, father = 4, noffs = 1, sex = 2)
CSF1PO <- marker(H1, alleles = c(10, 14, 15), 4, c(14, 15),
                 5, c(10, 14), 6, c(10, 15))
D7S820 <- marker(H1, alleles = c(11, 12), 4, c(11, 12),
                 5, c(11, 12), 6, c(11, 12))
```

```
D19S433 <- marker(H1, alleles = c(14, 99), 4, c(14, 14),
                    5, c(14, 14), 6, c(14, 14))
id <- c ("GF", "GM", "Brother", "Defendant", "Mother",
          "Child")
plot(H1,list(CSF1PO,D7S820,D19S433), id.labels = id,
     title = ("My brother did it! CSF1PO, D7S820, D19S433"))
```

Exercise 5.6 Exclusion probability. Below we explain the calculations leading to Equation 2.15 with the function `exclusionPower` of the `R` package `paramlink` (see the help pages for further details).

(a) Show that the probability of excluding a man in a duo case for the marker D3S1358 is found by entering

```
require(paramlink)
require(Familias)
data(NorwegianFrequencies)
p <- NorwegianFrequencies$D3S1358
claim <-  nuclearPed(noffs = 1, sex = 1)
true <- list(singleton(id = 1,sex = 1),
              singleton(id = 3, sex = 2))
available <- c(1, 3)
PE1 <- exclusionPower(claim, true, available,
                       alleles = length(p), afreq = p)
```

(b) Find the exclusion power for TPOX and verify Equation 2.15.

Exercise 5.7 Y haplotype evidence. `disclapmix`. Below we step through the note "Tutorial on haplotype evidence with focus on Y-STR haplotype analysis using the discrete Laplace method" by Mikkel Meyer Andersen.[4] It is a challenging problem to estimate the match probability if the evidence obtained (in this case a Y haplotype) has not been observed in the relevant database. Some alternatives are explored below.

(a) Load the package `disclap` and look at the database `danes`:

```
require(disclapmix)
data(danes)
str(danes)
```

(b) Estimate the match probability by $1/(N+1)$, where N is the number of haplotypes in the database. Save your answer in `p.count`.

(c) Find the number of singletons s in the database—that is, the number of haplotypes occurring once. Estimate the match probability with Brenner's formula $(1-k)/(N+1)$, where $k=(s+1)/(N+1)$, and save the answer as `p.brenner`.

[4] http://arken.umb.no/~theg/Copenhagen2014/tutorial.pdf.

(d) Estimate the match probability according to the Laplace approach by going through the steps below. First, these data are prepared a bit. The markers DYS389I and DYS389II are decoupled by setting DYS389II := DYS389II − DYS389I. Rather than keeping a record of the number of each haplotype (as recorded in `danes$n`); the transformed database `danes_db` contains copies of identical haplotypes. The above is achieved by[5]

```
danes_cor <- danes
danes_cor$DYS389II <- with(danes_cor, DYS389II - DYS389I)
danes_db <- as.matrix(danes_cor[ rep(1L : nrow(danes_cor) ,
                        danes_cor$n), 1L : 10L])
```

Check that the number of lines in the new version of the database, `danes_db`, equals 185.

(e) The previously mentioned tutorial shows that the model based on four clusters of haplotypes is optimal. This Laplace model is fitted below:

```
fit <- disclapmix(x = danes_db, clusters = 4L,
                  iterations = 500L)
```

The singletons are

```
singletons <- as.matrix(subset(danes_cor,
                        n == 1L)[, 1L : 10L])
```

and the match probabilities for these singletons are

```
p.disclap <- predict(fit, newdata = singletons )
```

Find the smallest and the largest match probability.

(f) Remove the first singleton in the database. Refit the model, and estimate the match probability for the removed haplotype.

Exercise 5.8 Estimation frequencies. `DNAtools`. Estimation of allele frequencies and simulation of databases based on the R package `DNAtools` are presented below.

(a) Load and look at the database `dbExample` by typing

```
require(DNAtools)
data(dbExample)
head(dbExample, 5)  # prints first 5 lines
tail(dbExample, 5)  # prints last 5 lines
str(dbExample)      # displays the structure of the database
```

[5]The notation `1L` below is used to indicate that 1 is to be interpreted as an integer.

Estimate and save the allele frequencies by entering

```
phat <- freqEst(dbExample)
```

(b) Plot the allele frequencies for the first marker:

```
barplot(phat[[ 1 ]], sub = names(phat.new)[[1]])
```

Plot the allele frequencies for all markers, one at a time:

```
par(ask = TRUE)
lapply(phat, barplot)
par(ask = FALSE)  #  Turns of pausing
```

(c) Use the function `dbSimulate` to simulate a database of the same size as `dbExample` based on the estimate `phat`, and reestimate the allele frequencies. Save the new estimates as `phat.new` and plot them as for `phat` above.

CHAPTER

Models for pedigree inference

6

CHAPTER OUTLINE

In this chapter, we restate our general framework for pedigree inference in a somewhat more theoretical language than before. The purpose is to give a better overview of possible models and how computations within these models can be done. Chapters 7 and 8 then focus on uncertainty in the parameters of our models and the use of models to make decisions, respectively.

Our basic assumption is that we have DNA data $\mathcal{D}$ for a set of persons (or possibly animals or plants), and from this we would like to learn about how the persons are related. We assume that we are given a number of possible pedigrees $\mathcal{P}_1, \mathcal{P}_2, \ldots, \mathcal{P}_n$ relating the persons, and we use the data $\mathcal{D}$ to help choose between these. An alternative is to try to construct possible pedigrees from the genetic data; this approach is, however, not covered in this book. As all humans are in fact related in one big pedigree, the specification of possible pedigrees $\mathcal{P}_1, \dots, \mathcal{P}_n$ not only restricts what we regard as possible but also formulates each possibility in terms of a specific pedigree and an assumption that its *founders*—that is, those persons in the pedigree who do not have their parents in the pedigree—are "unrelated." Thus, we need some kind of *population model* to specify what "unrelated" means.

An important final ingredient in our standard setup is a set of prior probabilities $p_1, \dots, p_n$ for the pedigrees $\mathcal{P}_1, \dots, \mathcal{P}_n$. With these specified, we use Bayes's formula to compute the posterior probability for each pedigree $\mathcal{P}_j$ given the data $\mathcal{D}$:

$$\Pr(\mathcal{P}_j \mid \mathcal{D}) = \frac{p_j \Pr(\mathcal{D} \mid \mathcal{P}_j)}{\sum_{k=1}^{n} p_k \Pr(\mathcal{D} \mid \mathcal{P}_k)}.$$

We need to specify a model to compute the probability $\Pr(\mathcal{D} \mid \mathcal{P})$ for each possible pedigree $\mathcal{P}$. If we write $\mathcal{F}$ for the genotypes of the founders of the pedigree and $\mathcal{G}$ for the genotypes of the DNA-tested persons, we may subdivide such a model into three parts as shown in Figure 6.1. A model for the genotypes $\mathcal{F}$ of the founders, which in the following we will call a *population-level model*, a model for the genotypes $\mathcal{G}$ of the tested persons given the genotypes $\mathcal{F}$ and the pedigree $\mathcal{P}$, which we will call a *pedigree-level model*, and finally a model for the data $\mathcal{D}$ given the genotypes $\mathcal{G}$ of the DNA-tested persons, which we will call an *observational-level model*. The probability of the data given each pedigree, $\Pr(\mathcal{D} \mid \mathcal{P})$, can then be expressed as

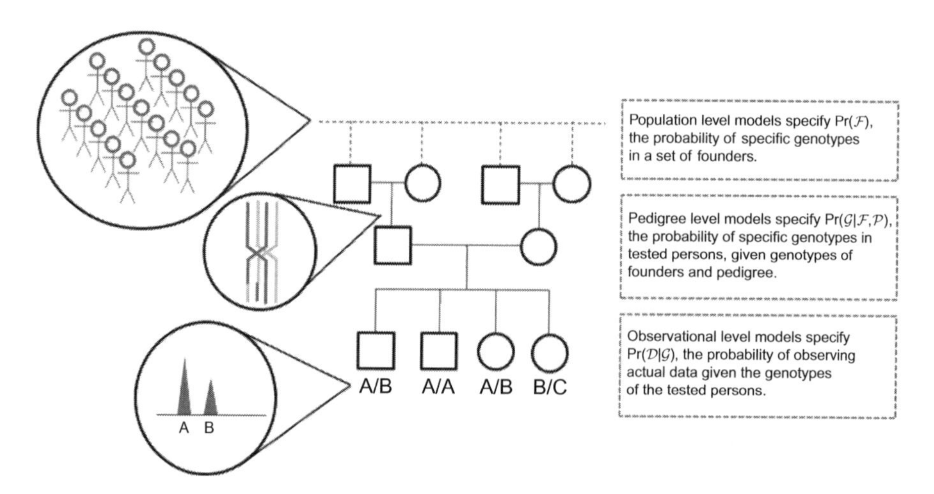

FIGURE 6.1

The three-level concept. The top level concerns the population of the founders, the middle level the inheritance within the pedigree, and the bottom level differences between the actual observations and the phased genotypes of tested persons.

$$\Pr(\mathcal{D} \mid \mathcal{P}) = \sum_{\mathcal{G}} \sum_{\mathcal{F}} \Pr(\mathcal{D}, G, F \mid \mathcal{P})$$
$$= \sum_{\mathcal{G}} \sum_{\mathcal{F}} \Pr(\mathcal{D} \mid G) \Pr(G \mid F, \mathcal{P}) \Pr(F), \qquad (6.1)$$

where the sums go through all possible values of $\mathcal{G}$ and $\mathcal{F}$. In the last equation, we use some reasonable conditional independence assumptions: We assume that, given the genotypes of the tested persons, the distribution of the actual observed data is not influenced by the pedigree relating the persons or the genotypes of (untyped) founders of the pedigree. In other words, we include in $\mathcal{G}$ any information that could possibly influence the observed data. We also assume that information about what pedigree the founders are part of is independent of their genotypes.

Actual computations will usually be done in a smarter way than going through the often large sums in Equation 6.1. In Section 6.4, we will summarize some of the commonest ways to achieve computational efficiency. But until then, our focus will be on defining different variants of population-level models $\Pr(\mathcal{F})$, pedigree-level models $\Pr(\mathcal{G} \mid \mathcal{F}, \mathcal{P})$, and observational-level models $\Pr(\mathcal{D} \mid \mathcal{G})$.

As our initial focus is to describe models, we will use a notation that facilitates such descriptions: By $\mathcal{G}$ we mean the *phased genotype* data of the DNA-tested persons. In other words, we include in $\mathcal{G}$ information about whether each allele is maternal or paternal, even if such information is generally not easily observable. Similarly, $\mathcal{F}$ indicates the phased genotypes of the founders. In fact, we allow pedigrees with persons having one parent but not the other in the pedigree. This differs from the traditional formulations whereby both or no parents are assumed to be present in the pedigree; see, for instance, [90]. Thus, $\mathcal{F}$ will actually describe the ordered list of *founder alleles* or *founder haplotypes*: these include the two alleles (or haplotypes) of persons who are founders, but they also include the paternal allele (or haplotype) of persons whose father is not in the pedigree but whose mother is, and vice versa.

Often we will have a sequence of m markers, in which case we will use notation such as $\mathcal{D}_i$, $\mathcal{G}_i$, and $\mathcal{F}_i$ for $i = 1, \ldots, m$ to indicate how these variables are split up into marker-specific information.

For some alternative ways to summarize models for pedigree inference, see [93–96].

6.1 POPULATION-LEVEL MODELS

We start with models for $\Pr(\mathcal{F})$—that is, models which specify the probability of a specific sequence $\mathcal{F} = (f_1, f_2, \ldots, f_F)$ of founder alleles or founder haplotypes. We will first assume we are considering a single autosomal marker; we will return to the multiple-marker situation in Section 6.1.4.

Our main model so far has been the following: There is a set of k alleles considered possible, and they are observed with probabilities, or *allele frequencies*, which may be specified in some vector $\mathbf{p} = (p_1, p_2, \ldots, p_k)$ of fixed and known positive values summing to 1. The different alleles observed in different persons, or the same person, are independent, and the probability that each is of type i is given by p_i. A well-known consequence of this model is that of Hardy-Weinberg equilibrium, in which the probability of observing a genotype A/B is $2p_A p_B$ when A and B are different and p_A^2 when A and B are the same. However, observations of real data may easily indicate small or large departures from such probabilities. One may also observe indications that allele observations between unrelated persons, or between different loci, are dependent. Thus, there is a need for more refined models.

In terms of biology, there are many reasons why the simplistic model above would be insufficient. It may be derived by our assuming there is a fixed and unchanged set of alleles, and that, in each generation, there is "random mating" so the paternal allele in a person has equal probability of coming from any of the males in the previous generation, and *independently of this*, the maternal allele has equal probability of coming from any of the females. Clearly, mating cannot really be described as random in any natural population. More importantly, the set of alleles is not fixed and unchanged, but is evolving all the time. Mutations occur, changing allele frequencies over time. When a mutation occurs, the new allele will initially be inherited together with the alleles close to it on the same chromosome, until crossovers slowly reduce this dependency. This dynamic development together with the lack of randomness in mating creates very complex dependencies not easily described with simple mathematical models. In this book, we will just mention some of the commonest extensions of the basic model above. We start by looking at consequences of uncertainties in allele frequencies. Although this discussion could have been placed in the next chapter, where the focus is on parameter uncertainty, it is placed in this chapter, as the issue is closely connected with common ways of dealing with subpopulations in population-level models.

6.1.1 *FREQUENCY UNCERTAINTY

A "frequency" is by definition computed from a set of observations as the count of observations of a certain type divided by the total number of observations. The most basic procedure for finding allele frequencies does just this: A database of persons is collected, and one assumes both alleles observed for each person are randomly sampled from a population of alleles. Let us write $\mathbf{C} = (C_1, \ldots, C_v)$ for the vector of counts of observations for the v different alleles observed, and let N be the total number of alleles counted (so N is twice the number of persons in the database, when the marker is autosomal). Then one may use as parameters in our computations the allele frequencies derived by dividing the counts C_i by N. In vector form, we may write that the vector $\mathbf{p}$ of allele frequency parameters is computed as

$$\mathbf{p} = \mathbf{C}/N.$$

The model parameter p_i is supposed to represent the probability that an allele in a pedigree founder has type i. Estimating this probability with a frequency is not necessarily optimal. The biggest problem occurs when case data contains alleles that have not been observed in the database. The computed frequency is thus zero.

However, as the allele is observed in the case data, it becomes meaningless to use a model where the probability of observing it is zero. The probability should instead be some small positive number for any allele that is considered possible on the basis of knowledge of molecular biology. The number should be small, as we have not yet seen the allele, but it should be positive.

We will continue to call the components of the vector **p** allele "frequencies," although "allele probabilities" would have been a better name when we view each of them as indicating the probability that the next allele we observe will be of a certain type. However, the term "allele frequency" is well established, and we stick to this nomenclature.

To obtain a better estimate for **p** than $\mathbf{C}/N$, we apply a Bayesian perspective. We assume there are k possible alleles: this set of possible alleles includes those observed in the database, but may also include others. For example, if we have observed at a short-tandem repeat (STR) marker alleles of lengths 12, 13, 15, 16, and 17, it is reasonable to assume that 14 is also a possible length. As mutations to alleles of fractional lengths (so-called microvariants) happen, one should also consider such alleles as being possible, but even less probable before we have seen them than unseen alleles representing a full number of replications.

Let us rewrite the count vector **C** as a vector over the k possible alleles, so that it may have components that are zero. Its sampling distribution given **p** can then be modeled as multinomial:

$$\Pr(\mathbf{C} \mid \mathbf{p}) = \frac{N!}{C_1!C_2!\dots C_k!} p_1^{C_1} p_2^{C_2} \cdots p_k^{C_k}.$$

The natural conjugate prior to use when learning about **p** from such multinomial data is the Dirichlet distribution. This distribution has been used by several authors in this context. See, for example, [97], where forensic match probabilities are modeled starting with the same prior.

A vector **p** with elements that sum to 1 has a Dirichlet distribution with parameter $\lambda = (\lambda_1, \dots, \lambda_k)$, $\lambda_i > 0$, when the probability density is

$$\pi(\mathbf{p} \mid \lambda) = \frac{1}{B(\lambda)} \prod_{i=1}^{k} p_i^{\lambda_i - 1}, \tag{6.2}$$

where $B(\lambda)$ is the multivariate beta function:

$$B(\lambda) = \frac{\prod_{i=1}^{k} \Gamma(\lambda_i)}{\Gamma(\sum_{i=1}^{k} \lambda_i)}.$$

Here, the function $\Gamma(t)$ is the gamma function[1] defined by

$$\Gamma(t) = \int_0^\infty x^{t-1} e^{-x}\, dx.$$

[1]For the gamma function, note that when t is a positive integer, $\Gamma(t) = (t-1)!$, and more generally we have $\Gamma(t+k)/\Gamma(t) = t \times (t+1) \times (t+2) \times \cdots \times (t+k-1)$.

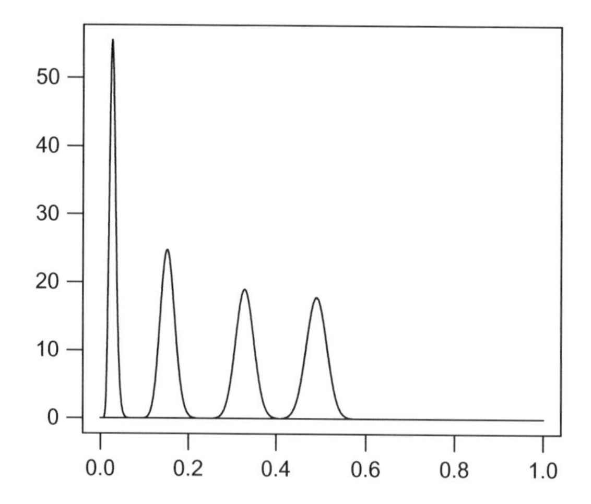

FIGURE 6.2

The marginal beta distributions of the components of a vector with a Dirichlet distribution with parameter vector (14, 245, 77, 164).

Note that the marginal distribution for each p_i is a beta distribution with parameters λ_i and $\Lambda - \lambda_i$, where we write $\Lambda = \lambda_1 + \lambda_2 + \cdots + \lambda_k$. Note also that the variance of this beta distribution is $\frac{\lambda_i(\Lambda-\lambda_i)}{\Lambda^2(\Lambda+1)}$, so the variance decreases when Λ increases. In other words, the uncertainty of the vector $\mathbf{p}$ decreases as Λ increases. Figure 6.2 shows the marginal beta distributions of $\mathbf{p} = (p_1, p_2, p_3, p_4)$ when it has a Dirichlet distribution with parameter $\lambda = (14, 245, 77, 164)$.

If the prior for $\mathbf{p}$ is Dirichlet with parameter λ, we get that the posterior for $\mathbf{p}$ after observing $\mathbf{C}$ is also Dirichlet, in fact with parameter vector $\mathbf{C} + \lambda$, as seen below:[2]

$$\pi(\mathbf{p} \mid \mathbf{C}) \propto_{\mathbf{p}} \Pr(\mathbf{C} \mid \mathbf{p})\pi(\mathbf{p}) = \frac{N!}{C_1!C_2!\cdots C_k!} p_1^{C_1} p_2^{C_2} \cdots p_k^{C_k} \frac{1}{B(\lambda)} \prod_{i=1}^{k} p_i^{\lambda_i - 1} \tag{6.3}$$

$$\propto_{\mathbf{p}} \prod_{i=1}^{k} p_i^{\lambda_i + C_i - 1}.$$

The sum of the parameters of the posterior distribution is $S = N + \Lambda$. So when the database increases in size—that is, when N increases, the uncertainty about the allele frequencies decreases.

If we want to obtain a particular estimate for $\mathbf{p}$, we can use the posterior expectation. The expectation of a Dirichlet distribution with parameter vector λ is the vector λ/Λ. Thus, the expectation of the posterior is

$$\mathrm{E}(\mathbf{p} \mid \mathbf{C}) = \frac{\lambda + \mathbf{C}}{\Lambda + N}.$$

[2]The notation $\propto_{\mathbf{p}}$ means proportional as a function of $\mathbf{p}$.

If, for example, we start with a prior where $\lambda_i = 1$ for all i, we get $\Lambda = k$, and $E(p_i) = \frac{1+C_i}{k+N}$. For an allele which is not observed in the database, so $C_i = 0$, we get $E(p_i) = \frac{1}{k+N}$, so a small positive probability is assigned to such an allele even if it has not been observed. The numbers λ_i are sometimes called *pseudocounts* as they fill the same function as the actual counts C_i without being counts. They do not even have to be whole numbers, and they can be different for different alleles. For example, microvariant alleles can be assigned a smaller pseudocount than regular alleles. If all the pseudocounts approach zero, we get back the traditional frequentist estimate of allele frequencies as $\mathbf{C}/N$.

Example 6.1 Unseen alleles. To do likelihood computations with case data containing alleles that have not been seen before in the database, different solutions have been proposed in the literature. Some adjustments to avoid the unfortunate estimate of zero are discussed in [14, Sections 6.4 and 6.5] and [13, Section 6.3.1]. One proposal is to use the allele frequency $5/N$ for such alleles as explained in [98, p. 477] (remember that our N is defined as two times the number of people in the database). One problem with this is that the allele frequencies no longer sum to 1. The proposal is somewhat similar to the use of pseudocounts of 5 for all possible alleles, although the latter option leads to allele frequencies of $5/(5k + N)$ for alleles not in the database. A reason for the use of 5 instead of 1 may be that one wants results to be "conservative": very small p_i may lead to very large or very small likelihood ratios (LRs), and one does not want such LRs to depend too much on parameters for which the estimation procedure is debatable.

6.1.2 *TAKING FREQUENCY UNCERTAINTY INTO ACCOUNT

Any likelihood is computed from some model with specific parameters. If θ is one such parameter, we can use an estimation procedure (as in the previous section) to obtain an estimate $\hat{\theta}$ to be used in the computation. However, we may worry that the uncertainty in this estimate may influence our likelihood computation in a biased, nonlinear way.

The likelihood is by definition the probability of data given the model and the hypothesis H. This means that if the model depends on a parameter θ for which we have a probability distribution, we need to integrate it out to compute the likelihood:

$$\Pr(\text{data} \mid H) = \int_\theta \Pr(\text{data}, \theta \mid H)\, d\theta = \int_\theta \Pr(\text{data} \mid \theta, H)\pi(\theta)\, d\theta.$$

Here, $\Pr(\text{data} \mid \theta, H)$ is the probability for the data for a given θ, while $\pi(\theta)$ is the probability density[3] reflecting our knowledge of the parameter, in practice the density of a posterior distribution like the one derived in the previous section. Computing the

[3]If θ is a discrete parameter, one should use the probabilities $\Pr(\theta)$ and a sum instead of an integral.

integral will take into account the uncertainty in θ, and will generally give a different value for the likelihood compared with computing

$$\Pr(\text{data} \mid H) \approx \Pr(\text{data} \mid \hat{\theta}, H).$$

In our particular case, note first that if the parameter $\mathbf{p}$ has a Dirichlet distribution with parameter λ, and for any vector $\mathbf{c} = (c_1, \ldots, c_k)$, where for each i, $c_i > -\lambda_i$, it follows from the definition of the Dirichlet distribution, Equation 6.2, that

$$\int_{\mathbf{p}} \prod_{i=1}^{k} p_i^{c_i} \pi(\mathbf{p})\, d\mathbf{p} = \frac{B(\lambda + \mathbf{c})}{B(\lambda)}. \tag{6.4}$$

Assume $\mathbf{c} = (c_1, \ldots, c_k)$ represents the counts of the various alleles that occur in the ordered sequence $\mathcal{F} = (f_1, \ldots, f_F)$, with the sum of the c_i equal to n. Then,

$$\Pr(\mathcal{F} \mid \mathbf{p}) = \prod_{i=1}^{k} p_i^{c_i}.$$

Thus, we get

$$\Pr(\mathcal{F}) = \int_{\mathbf{p}} \Pr(\mathcal{F} \mid \mathbf{p}) \pi(\mathbf{p})\, d\mathbf{p} = \frac{B(\lambda + \mathbf{c})}{B(\lambda)}. \tag{6.5}$$

As we saw in the previous section, when the distribution of $\mathbf{p}$ is based on a database with counts $\mathbf{C}$, it has a Dirichlet distribution with parameter $\lambda + \mathbf{C}$, where λ represents the pseudocounts, so we get $\Pr(\mathcal{F}) = B(\lambda + \mathbf{C} + \mathbf{c})/B(\lambda + \mathbf{C})$. In fact we may rewrite this as follows:

$$\begin{aligned}
\Pr(\mathcal{F}) &= \frac{B(\lambda + \mathbf{C} + \mathbf{c})}{B(\lambda + \mathbf{C})} \\
&= \frac{\prod_{i=1}^{k} \Gamma(\lambda_i + C_i + c_i) \times \Gamma(\Lambda + N)}{\Gamma(\Lambda + N + n) \times \prod_{i=1}^{k} \Gamma(\lambda_i + C_i)} \\
&= \frac{\Gamma(\Lambda + N)}{\Gamma(\Lambda + N + n)} \times \prod_{i=1}^{k} \frac{\Gamma(\lambda_i + C_i + c_i)}{\Gamma(\lambda_i + C_i)} \\
&= \frac{\prod_{i=1}^{k} [(\lambda_i + C_i)(\lambda_i + C_i + 1) \cdots (\lambda_i + C_i + c_i - 1)]}{(\Lambda + N)(\Lambda + N + 1) \cdots (\Lambda + N + n - 1)} \\
&= \prod_{j=1}^{n} \frac{\lambda_{a_j} + C_{a_j} + b_j}{\Lambda + N + j - 1}.
\end{aligned} \tag{6.6}$$

For each of the $j = 1, \ldots, n$ alleles in the ordered sequence $\mathcal{F}$, we use here a_j to denote allele j and b_j to denote the count of such alleles before the jth place in the sequence.

There is a nice interpretation of the formula above. When in a likelihood computation we want to compute the probability of observing the sequence $\mathcal{F}$ of founder alleles in a pedigree, we, in general, compute this as a sequence of conditional probabilities: The probability of observing the first allele given knowledge obtained from our database times the probability of observing the second allele given knowledge from the database *and* the fact that we have observed the first allele times the corresponding conditional probability for the third allele, and so on. If we believe the case data alleles are sampled from the same population as the database alleles, we should simply add the count of the first allele to the database before computing the second probability, add the counts of the two first alleles before computing the third probability, and so on. Using the expectations found in the previous section, we get for the first allele

$$\frac{\lambda_{a_1} + C_{a_1}}{\Lambda + N}.$$

If the second allele is a different one—that is, if $a_1 \neq a_2$, we get the product

$$\frac{\lambda_{a_1} + C_{a_1}}{\Lambda + N} \times \frac{\lambda_{a_2} + C_{a_2}}{\Lambda + N + 1},$$

whereas if $a_1 = a_2$, we get

$$\frac{\lambda_{a_1} + C_{a_1}}{\Lambda + N} \times \frac{\lambda_{a_2} + C_{a_2} + 1}{\Lambda + N + 1}.$$

In fact, we can see that the whole product becomes equal to the formula derived above. So taking the uncertainty of the allele frequencies into account is equivalent to using allele frequencies computed in a way that includes the alleles in the founders of the pedigree sequentially into the database.

The niceness of the computational results above depends on the use a prior that is conjugate to the multinomial sampling distribution. See, for example, [94] for a similar exposition where beta priors are used for allele frequencies.

6.1.3 *POPULATION STRUCTURE AND SUBPOPULATIONS

The idea that the alleles in the founders of a pedigree are randomly sampled from one population, with allele frequencies **p**, is obviously a simplification. It assumes that when people have children, they are equally likely to do so with any person of the opposite sex in that population, and will not have children with any person outside what is defined as their population. The reality is that people are most likely to produce offspring with persons who live reasonably close to them and have a similar socioeconomic and cultural background, and that the probability to produce children with somebody will decrease gradually as such similarities decrease. Modeling this is, of course, very difficult, in particular as the degree of mixing of people with different backgrounds has increased over time. In this book, we will consider only the simplest (and most used) way to model the effect described above.

Assume the founders of the pedigree come from some subpopulation for which we have no database. We should then use Equation 6.6 with $N = 0$ and each $C_i = 0$ to compute $\Pr(\mathcal{F})$. However, one might know that the subpopulation is fairly similar to one or more other populations, for which databases exist, and this knowledge can then be used to construct a sensible prior vector $\lambda = (\lambda_1, \ldots, \lambda_k)$. For example, if we have three populations A, B, and C, with which the subpopulation has similarities, we might use

$$\lambda = w_{\mathrm{A}}(\lambda_{\mathrm{A}} + C_{\mathrm{A}}) + w_{\mathrm{B}}(\lambda_{\mathrm{B}} + C_{\mathrm{B}}) + w_{\mathrm{C}}(\lambda_{\mathrm{C}} + C_{\mathrm{C}}),$$

where, for example, λ_{A} and C_{A} are the prior vector and count vector for population A, and where w_{A}, w_{B}, and w_{C} are weights indicating how close each population is to the subpopulation. For example, if such a weight is 1, this indicates that each person in the A database is a random sample from the subpopulation we are studying. If, on the other hand, the weight is 0.01, this indicates that only 1% of the people in the A database can be assumed to be randomly sampled from the subpopulation. So, in general, the weights must be smaller than 1, and are usually quite small.

Let us now instead use the perspective that there is one population that is similar to the subpopulation the founders come from, and that we have enough data for the former population to fix its vector of allele frequencies $\hat{\mathbf{q}}$ with ignorable uncertainty. Let us define the vector λ for the Dirichlet prior distribution for the subpopulation frequencies as

$$\lambda = (1/\theta - 1)\hat{\mathbf{q}}, \tag{6.7}$$

where $0 < \theta < 1$ is some positive number. When θ is very small, the sum of λ, which is $1/\theta - 1$, will be very large, indicating the allele frequencies of the subpopulation are very close to $\hat{\mathbf{q}}$. When θ instead is large, $1/\theta - 1$ will be small, indicating the subpopulation is quite separate; in other words, people tend to mate within the subpopulation. Applying Equation 6.6, we get

$$\Pr(\mathcal{F}) = \prod_{j=1}^{n} \frac{(1/\theta - 1)\hat{q}_{a_j} + b_j}{(1/\theta - 1) + j - 1} = \prod_{j=1}^{n} \frac{(1-\theta)\hat{q}_{a_j} + b_j\theta}{(1-\theta) + (j-1)\theta} = \prod_{j=1}^{n} \frac{b_j\theta + (1-\theta)\hat{q}_{a_j}}{1 + (j-2)\theta}.$$

We see that we get the sampling formula of Equation 2.13. In other words, using a kinship parameter θ to represent the degree of separation of a subpopulation corresponds to using a Dirichlet distribution for the subpopulation frequencies with the parameter given by Equation 6.7.

The connection between allele frequency uncertainty and the θ correction can be used in several ways. For example, many available types of software for relationship computations include the possibility to use a θ correction. If one wants to take allele frequency uncertainty into account by using a Dirichlet distribution with parameter

$\Lambda\hat{\mathbf{q}}$ for the allele frequencies, Equation 6.7 shows that one can do this by using a θ correction with

$$\theta = \frac{1}{\Lambda + 1}.$$

This may be important when one is using a small database. The connection may also be used in the other direction: Suppose in a particular case we would like to compute a likelihood $\Pr(\mathcal{D} \mid H, \theta = \hat{\theta})$ using a nonzero value $\hat{\theta}$ for the θ correction. We have

$$\Pr(\mathcal{D} \mid H, \theta = \hat{\theta}) = \sum_{\mathcal{F}} \Pr(\mathcal{D} \mid \mathcal{F}, H) \Pr(\mathcal{F} \mid \theta = \hat{\theta}),$$

and using Section 6.1.2 and Equation 6.7,

$$\Pr(\mathcal{F} \mid \theta = \hat{\theta}) = \int_{\mathbf{p}} \Pr(\mathcal{F} \mid \mathbf{p}) \pi(\mathbf{p}) \, d\mathbf{p},$$

when $\pi(\mathbf{p})$ is a Dirichlet density with parameter $(1/\theta - 1)\hat{\mathbf{q}}$. If we simulate R vectors of allele frequencies $\mathbf{p}_1, \mathbf{p}_2, \ldots, \mathbf{p}_R$ from this distribution, we can approximate the integral above as

$$\Pr(\mathcal{F} \mid \theta = \hat{\theta}) \approx \frac{1}{R} \sum_{j=1}^{R} \Pr(\mathcal{F} \mid \mathbf{p}_j),$$

and we get

$$\begin{aligned} \Pr(\mathcal{D} \mid H, \theta = \hat{\theta}) &\approx \sum_{\mathcal{F}} \Pr(\mathcal{D} \mid \mathcal{F}, H) \times \frac{1}{R} \sum_{j=1}^{R} \Pr(\mathcal{F} \mid \mathbf{p}_j) \\ &= \frac{1}{R} \sum_{j=1}^{R} \Pr(\mathcal{D} \mid \mathbf{p}_j, H). \end{aligned}$$

Thus, instead of computing with a nonzero θ correction, we can simulate allele frequencies as above and average over the results. This approximation may be useful if use of the θ correction is computationally difficult for some reason.

Occasionally, the set of founders in a pedigree may be subdivided into subsets so that each subset comes from one specific population. Computations for this type of extension present no theoretical difficulties as long as we assume it is known which population each founder belongs to, and that the populations are independent. For each population, we may use the Dirichlet model above. The likelihood for the founders' genotypes simply becomes the product over the founders in each population.

More realistically, it is not completely known for all founders which population they belong to. This can still be handled if we have a list of possible assignments to populations, and prior probabilities for each of these assignments. Note, however, that summing over the possible assignments must be done when computing the probability for each possible combination of founder haplotypes, and should not be done separately for each marker, even if the markers are otherwise assumed to be independent. The reason is that uncertainty in population assignment induces dependencies between the markers.

6.1.4 *HAPLOTYPE MODELS

The number of possible haplotypes covering a sequence of m markers may be huge, in particular when there are many possible alleles at each marker. Estimating the probability of observing a haplotype as the frequency with which it has been observed in a database becomes even more problematic than for single markers, as many haplotypes may be observed only once, if at all.

The simplest solution is to assume independence between markers: For the founder haplotypes $\mathcal{F} = (f_1, \ldots, f_F)$ we write $f_j = (f_{j1}, \ldots, f_{jm})$, where f_{ji} indicates the allele of founder haplotype j at marker i. Writing $p(i, a)$ for the probability of observing allele a at marker i, we may express the independence model (linkage equilibrium) as

$$\Pr(\mathcal{F}) = \prod_{j=1}^{F} \prod_{i=1}^{m} p(i, f_{ji}). \tag{6.8}$$

In our initial models, we have assumed that the probabilities $p(i, a)$ are fixed and known. It is possible to model the vector of probabilities for each marker with an independent Dirichlet distribution, and thus use a theta correction for each marker. Note, however, that this requires a known and fixed value for θ—characterizing the relevancy of the databases used for the subpopulation of the case data. If the degree of separateness of the subpopulation is not fixed and known, a dependency between markers may be introduced. For example, observing in the case data many copies of a rare allele at one marker indicates a high degree of subpopulation separateness, which may then increase the probability of observing repeated rare alleles also at other markers.

An alternative to Equation 6.8 is to apply to haplotypes the ideas applied to single markers in Sections 6.1.1 and 6.1.2. Specifically, if $\mathbf{p}$ is the vector of haplotype frequencies, we assume a Dirichlet prior for $\mathbf{p}$, which is then updated with counts from a database of haplotypes. If some separate population is available from which one may get estimates $\hat{q}(i, a)$ for allele frequencies for all markers i and all alleles a at each marker, we may use these to construct a prior: For all possible haplotypes $(i_1, \ldots, i_m)$, where i_j indicates the allele at marker j, define

$$\delta_{i_1,\ldots,i_m} = \lambda \times \hat{q}(1, i_1) \times \hat{q}(2, i_2) \times \cdots \times \hat{q}(m, i_m),$$

where λ is some positive number. We then use the vector δ of the $\delta_{i_1,\ldots,i_m}$ parameters as the parameter in the Dirichlet distribution. This prior takes as its starting point independency between markers, but acknowledges that actual haplotype frequencies may deviate from the haplotype frequencies expected under independence. The value of λ equals the sum of all the parameters $\delta_{i_1,\ldots,i_m}$, and thus it indicates the amount of uncertainty in the prior: a large λ indicates a belief that haplotype frequencies are close to those expected under independence (so there is little linkage disequilibrium), while a small λ gives a prior with larger probabilities for larger amounts of linkage disequilibrium.

If a vector of counts $\mathbf{c}$ of haplotypes is available from some database (ideally different from the one used to derive the values $\hat{q}(i,a)$) the posterior for the haplotype frequencies is a Dirichlet distribution with parameter vector $\delta + \mathbf{c}$. The posterior expected haplotype frequency of a haplotype $f_j = (f_{j1}, \ldots, f_{j,m})$ becomes

$$\frac{\delta_{f_{j1},\ldots,f_{j,m}} + c_{f_{j1},\ldots,f_{j,m}}}{\lambda + C},$$

where $c_{f_{j1},\ldots,f_{j,m}}$ is the count of the haplotype $(f_{j1}, \ldots, f_{j,m})$ and C is the total number of haplotypes in the database. This value is equal to the sampling formula in Equation 4.10. Disregarding uncertainty in the haplotype frequencies, one may compute $\Pr(\mathcal{F})$ as a product over such posterior expected frequencies:

$$\Pr(\mathcal{F}) = \prod_{j=1}^{F} \Pr(f_j) = \prod_{j=1}^{F} \frac{\delta_{f_{j1},\ldots,f_{j,m}} + c_{f_{j1},\ldots,f_{j,m}}}{\lambda + C}. \tag{6.9}$$

An alternative is to take into account the haplotype frequency uncertainty, in a similar way as Equation 6.5: Let $\mathbf{d}$ be the vector of counts of haplotypes in $\mathcal{F} = (f_1, \ldots, f_F)$. Then we get

$$\Pr(\mathcal{F}) = \frac{B(\delta + \mathbf{c} + \mathbf{d})}{B(\delta + \mathbf{c})}.$$

Note: This equation can be rewritten in the same way as Equation 6.6.

A way to describe the approach above is that it uses pseudocounts whose size is based on the allele frequencies of the alleles that are part of each haplotype. In that way, it might improve on a model with equal pseudocounts for all haplotypes. However, the resulting model for haplotype frequencies is not necessarily biologically realistic. The occurrence of recombinations means that a model where each haplotype is "sampled" from a database of "possible" haplotypes is not a natural fit. Another problem with the λ model discussed above is that it requires a value for the λ parameter. Haplotype frequencies, and thus LR results, may depend heavily on the value of λ, as discussed in Section 4.3. We return to the problem of estimating λ in Section 7.3.

An intermediary solution between modeling separate probabilities for all haplotypes and assuming independence between all markers is to use clusters of markers, where the markers within each cluster are treated as one haplotype marker, while one assumes independence between clusters. This may work well for clusters of single nucleotide polymorphism (SNP) markers, where the number of possible haplotypes within each cluster may be manageable.

Lastly we mention the possibility to use Markov models for haplotype probabilities. For a founder haplotype $f_j = (f_{j1}, \ldots, f_{j,m})$, it may be that there is correlation between allele types of markers that are close to each other in this sequence, and thus possibly close to each other on the chromosome, whereas alleles that are far away in the sequence are more or less independent. Probabilistically, we may assume, for example,

$$\Pr(f_{ji} \mid f_{j1}, \ldots, f_{j,i-1}) = \Pr(f_{ji} \mid f_{j,i-1})$$

for $i = 2, \ldots, m$, or more generally that

$$\Pr(f_{ji} \mid f_{j1}, \ldots, f_{j,i-1}) = \Pr(f_{ji} \mid f_{j,i-s}, \ldots, f_{j,i-1})$$

for $i = s+1, \ldots, k$ and some $s \geq 1$. In the first case, we say we use a Markov chain as a model, and in the second case an s-step Markov chain. With a Markov chain model, we need to specify parameters p_{1a} for the allele frequencies at the first marker, and then for $i = 2, \ldots, k$ *conditional probabilities* $p_{i,a|b}$—that is, the probabilities of observing allele a at marker i given that allele b has been observed at marker $i - 1$. With such probabilities defined, we may write

$$\Pr(\mathcal{F}) = \prod_{j=1}^{F} \left(p_{1,f_{j1}} \prod_{i=2}^{m} p_{i,f_{ji}|f_{j,i-1}} \right).$$

Markov models may be useful in some contexts, but often long-range dependencies between widely separated markers may be observed in databases. Such dependencies may in fact be a result of subpopulation structure, in other words a result of the reality of nonrandom mating. The best solution may then be to improve modeling of subpopulation structures. However, we do not discuss such models here.

6.1.5 POPULATION MODELS FOR NONAUTOSOMAL MARKERS

We started this chapter by stating that the model for the genotypes of the DNA-tested persons can be divided into a model for the genotypes of the founders of the pedigree, and a model for how these genotypes are inherited within the pedigree. For autosomal markers, it is in most cases natural to assume the founder genotypes are identically and independently distributed given the population frequencies of some population. The reason is that the dependency between autosomal genotypes diminishes very quickly when the relationship becomes more distant; there is rarely much information that can distinguish a pairwise relationship over more than five to six meioses from an "unrelated" relationship. The above statement is made precise and explored in [99].

This may, however, be different for other types of markers. As an example, Y-chromosomal markers are inherited from father to son mostly unchanged. So a finer subdivision of the concept of "unrelated" is needed in order to make inferences based on such data: men with ancestors who come from the same area are much likelier to have similar or identical Y chromosomes than men with ancestors from different areas.

The best way to deal with this may be to use not one population model, representing "unrelated" persons, but a range of different population models, representing various hypotheses of relatedness beyond relatedness specified in particular pedigrees. One would then need prior probabilities for each type of relatedness, just as we use prior probabilities for various pedigrees in our standard models, so that one could compute posterior probabilities using the population models. We will not discuss such possibilities further here. Gusmao et al. [100] present recommendations on the use of Y STRs in forensic settings.

A new type of marker has been introduced that could vastly improve the distinction of distantly related persons. So-called rapidly mutating Y STR markers have recently received increasing attention [101]. The mutation rate is commonly greater than 1%, and distant paternal relatives are expected to have different haplotypes.

Example 6.2 Illustration of the effect of rapidly mutating Y markers. Assume two males are found to share the same haplotype consisting of 10 markers. We do not discuss the estimation of frequencies for Y haplotypes (there are excellent discussions in, e.g., [53]), but rather assume the frequency is known and denoted by F_S. We now consider the following hypotheses:

H_1: The two males are related with n generations between them through a paternal lineage.
H_2: The two males are unrelated.

The LR comparing the data under the two hypotheses may now be formed as

$$\text{LR} = \frac{\Pr(\text{data} \mid H_1)}{\Pr(\text{data} \mid H_2)} = \frac{(1-\mu)^{10n}}{F_S},$$

where we disregard the possibility that several mutations have cancelled each other out along the lineage. We may further plot the LR as a function of n assuming $F_S = 0.01$ and using $\mu = 0.01$ for rapidly mutating markers and $\mu = 0.001$ for markers with a "normal" mutation rate.

Figure 6.3 shows the decrease of LR with the number of generations for different mutation rates of the markers. For rapidly mutating Y STRs, the probability that two individuals separated by 30 generations will share exactly the same haplotype identical by descent (IBD) is comparable to the population frequency of the haplotype.

We may further consider the example where the two males have identical haplotypes for all but one marker, where a one-step difference is observed. We denote the frequencies for the two observed haplotypes as F_{S1} and F_{S2}, respectively, and

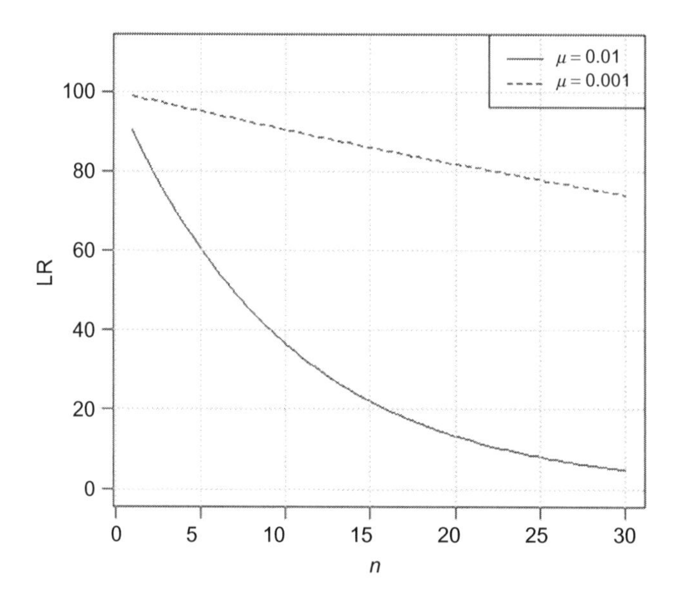

FIGURE 6.3

LR as a function of the number of generations for 10 Y STRs with different mutation rates.

assume that if a mutation happens, it adds or deletes one step with 50% chance. As a simplification, we disregard the possibility of multiple mutations. The probability that one of the observed haplotypes has transformed to the other after i generations is then $(1-\mu)^{9i} \times i\mu/2(1-\mu)^{i-1}$. We also make the simplifying assumption that H_1 means that the two males are both separated from their common paternal lineage ancestor by $n/2$ generations. By summing over the possible genotypes of the ancestor (s1 and s2), we obtain

$$\mathrm{LR} = \frac{\Pr(\text{data} \mid H_1)}{\Pr(\text{data} \mid H_2)} = \frac{(F_{\mathrm{S1}} + F_{\mathrm{S2}})(1-\mu)^{10n/2}(1-\mu)^{9n/2} \times n/2 \times \mu/2(1-\mu)^{n/2-1}}{F_{\mathrm{S1}} \times F_{\mathrm{S2}}}$$
$$= \frac{(F_{\mathrm{S1}} + F_{\mathrm{S2}})(1-\mu)^{10n-1} n\mu}{4F_{\mathrm{S1}}F_{\mathrm{S2}}}.$$

We may now further plot the LR as a function of the number of generations assuming $F_{\mathrm{S1}} = F_{\mathrm{S2}} = 0.01$ and using $\mu = 0.01$ for rapidly mutating markers and $\mu = 0.001$ for markers with a "normal" mutation rate.

The top around $n = 10$ of the LR curve in Figure 6.4 for rapidly mutating markers indicates how such alleles may be useful to discern paternal lineage relationships of this degree of closeness from both closer and more distant relationships.

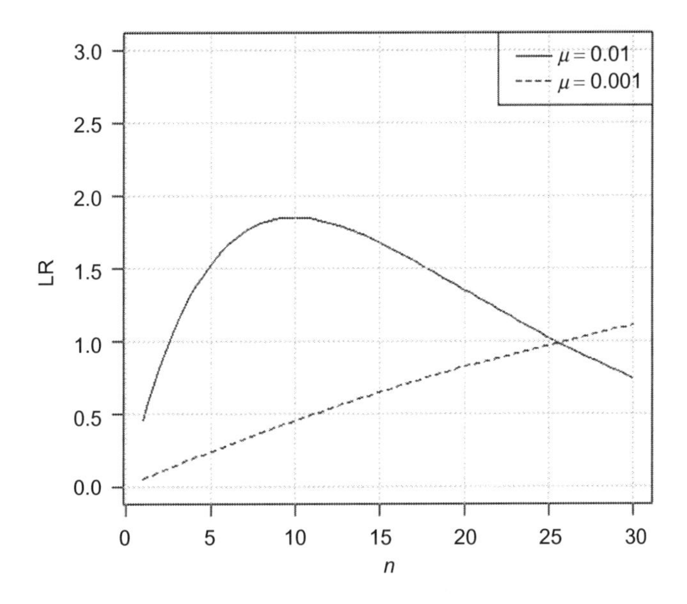

FIGURE 6.4

LR as a function of the number of generations in a paternal lineage separating two males, for 10 Y STRs with different mutation rates assuming a single one step difference in the data.

6.2 PEDIGREE-LEVEL MODELS

Given a pedigree and the phased genotypes $\mathcal{F}$ of its founders, what can we say about the phased genotypes $\mathcal{G}$ of the tested persons in the pedigree? Assume the pedigree contains R parent-child relationships. For $r = 1, \ldots, R$, let V_r be a variable describing whether the allele inherited from the parent (or, if several loci are considered, each inherited allele) comes from the paternal or the maternal chromosome of the parent. Then

$$\Pr(\mathcal{G}|\mathcal{F}) = \sum_{V_1,\ldots,V_R} \Pr(\mathcal{G}, V_1, \ldots, V_R \mid \mathcal{F})$$
$$= \sum_{V_1,\ldots,V_R} \Pr(\mathcal{G} \mid V_1, \ldots, V_R, \mathcal{F}) \prod_{r=1}^{R} \Pr(V_r). \tag{6.10}$$

The variables $V_1, \ldots, V_R$ together describe an *inheritance pattern*. Given this pattern and the phased founder genotypes, $\Pr(\mathcal{G} \mid V_1, \ldots, V_R, \mathcal{F})$ describes the probability of the genotype $\mathcal{G}$. If we assume that no mutations are possible, it will be either 0 or 1, depending on whether the inheritance pattern transfers the alleles of the founders given in $\mathcal{F}$ to the alleles of $\mathcal{G}$.

When we look at data from a single autosomal locus, V_r simply chooses the paternal or maternal allele at that locus. We will write $V_r = 1$ or $V_r = 0$ respectively in these two cases, and we have

$$\Pr(V_r = 1) = \Pr(V_r = 0) = \frac{1}{2}.$$

Thus, when we are considering data from a single autosomal locus i, we get

$$\Pr(\mathcal{G}_i|\mathcal{F}_i) = \sum_{V_1,\ldots,V_R} \Pr(\mathcal{G}_i \mid V_1,\ldots,V_R,\mathcal{F}_i)\frac{1}{2^R}.$$

For X-chromosomal loci, we have two possible values for every V_r corresponding to a mother-child relation, whereas V_r has only one possible value for father-daughter relations. The same is the case for all V_r when Y-chromosomal or mitochondrial alleles are transferred.

Example 6.3 Consider a pedigree with four persons: a child, a mother, a grandfather, and a grandmother. We have $R = 3$ parent-child relationships: mother-child (V_1), grandfather-mother (V_2), and grandmother-mother (V_3). There are five founder alleles at each locus: the four alleles of the grandfather and the grandmother, and the paternal allele of the child. Assume $\mathcal{F}_i = ((14,17),(14,17),14)$ describes the identities of these alleles (note that we use phased genotypes and that $(14,17)$ means that the paternal allele is 14 and the maternal allele is 17). If $\mathcal{G}_i = (14,14)$ is observed for the child for marker i, we can write

$$\Pr(\mathcal{G}_i|\mathcal{F}_i) = \sum_{V_1,V_2,V_3} \Pr(\mathcal{G}_i = (14,14) \mid V_1,V_2,V_3,\mathcal{F}_i = ((14,17),(14,17),14))\frac{1}{2^3}.$$

Going through the eight different combinations of values for V_1, V_2, and V_3, we see that exactly four of these ($V_1 = 1$, $V_2 = 1$, $V_3 = 1$; $V_1 = 1$, $V_2 = 1$, $V_3 = 0$; $V_1 = 0$, $V_2 = 0$, $V_3 = 1$; and $V_1 = 0$, $V_2 = 1$, $V_3 = 1$) result in a nonzero probability. Thus,

$$\Pr(\mathcal{G}_i|\mathcal{F}_i) = \frac{4}{8} = \frac{1}{2}.$$

We go on to study Equation 6.10 in the case where we have a sequence of m markers. Let us write

$$V_r = (v_{r1}, v_{r2}, \ldots, v_{rm}),$$

where each v_{ri} is either 0 or 1, indicating, respectively, that the maternal or paternal allele is inherited in relationship r at marker i. Let ρ_i denote the recombination rate between markers $i-1$ and i. According to the model for linkage discussed in Chapter 4, we have

$$\Pr(V_r) = q_{r1} \times q_{r2} \cdots q_{rm},$$

where $q_{r1} = 1/2$, while for $i > 1$

$$q_{ri} = \rho_i^{I(v_{ri} \neq v_{r,i-1})}(1-\rho_i)^{I(v_{ri} = v_{r,i-1})}.$$

In other words, for each marker $i = 2, 3, \dots, m$, we have a factor ρ_i if v_{ri} and $v_{r,i-1}$ are different and a factor $1 - \rho_i$ if they are equal. For the other part of Equation 6.10, we get

$$\Pr(\mathcal{G} \mid V_1, \dots, V_R, \mathcal{F}) = \Pr(\mathcal{G}_1, \dots, \mathcal{G}_m \mid V_1, \dots, V_R, \mathcal{F}_1, \dots, \mathcal{F}_m)$$
$$= \prod_{i=1}^{m} \Pr(\mathcal{G}_i \mid v_{1i}, \dots, v_{Ri}, \mathcal{F}_i).$$

From this we obtain a general equation for a pedigree-level model including linkage:

$$\Pr(\mathcal{G}|\mathcal{F}) = \sum_{V_1,\dots,V_R} \left(\prod_{i=1}^{m} \Pr(\mathcal{G}_i \mid v_{1i}, \dots, v_{Ri}, \mathcal{F}_i) \right) \prod_{r=1}^{R} \prod_{i=1}^{m} q_{ri}$$
$$= \sum_{V_1,\dots,V_R} \prod_{i=1}^{m} \left(\Pr(\mathcal{G}_i \mid v_{1i}, \dots, v_{Ri}, \mathcal{F}_i) \prod_{r=1}^{R} q_{ri} \right). \qquad (6.11)$$

The factor $\Pr(\mathcal{G}_i \mid v_{1i}, \dots, v_{Ri}, \mathcal{F}_i)$ in Equation 6.11 represents the probability of observing the phased genotype $\mathcal{G}_i$ in the tested persons given the phased genotype $\mathcal{F}_i$ of the founders and the inheritance pattern at locus i, described by $v_{1i}, \dots, v_{Ri}$. As mentioned before, if we assume that no mutations happen within the pedigree, this factor is either 0 or 1. However, in situations where we need to include mutations in our model, the factor becomes more complex.

Let $h_{1i}, \dots, h_{Hi}$ denote the alleles occurring in the pedigree at locus i that are not included in $\mathcal{G}_i$ or $\mathcal{F}_i$, and subdivide the alleles $\mathcal{G}_i$ into $\mathcal{G}_i^+$ and $\mathcal{G}_i^-$, where the first group are those alleles that are also in $\mathcal{F}_i$. Then we get

$$\Pr(\mathcal{G}_i \mid v_{1i}, \dots, v_{Ri}, \mathcal{F}_i) = \Pr(\mathcal{G}_i^+ \mid \mathcal{F}_i) \sum_{h_{1i},\dots,h_{Hi}} \Pr(\mathcal{G}_i^-, h_{1i}, \dots, h_{Hi} \mid v_{1i}, \dots, v_{Ri}, \mathcal{F}_i).$$

The factor $\Pr(\mathcal{G}_i^+ \mid \mathcal{F}_i)$ is clearly either 0 or 1. For the rth relationship in the pedigree, write c_{ri}, p_{ri}, and m_{ri} for the child, paternal, and maternal alleles of the relationship, respectively, at marker i. Then we get

$$\Pr(\mathcal{G}_i^-, h_{1i}, \dots, h_{Hi} \mid v_{1i}, \dots, v_{Ri}, \mathcal{F}_i) = \prod_{r=1}^{R} \Pr(c_{ri} \mid p_{ri}, m_{ri}, v_{ri})$$

as $\mathcal{F}_i$ contains exactly the founder alleles, and all the other alleles are listed as $h_{1i}, \dots, h_{Hi}$ and $\mathcal{G}_i^-$. So in order to complete a specification of a model including mutations, we must specify the probabilities

$$\Pr(c_{ri} \mid p_{ri}, m_{ri}, v_{ri}),$$

in other words, the probability that p_{ri} is transferred as c_{ri} when $v_{ri} = 1$ and the probability that m_{ri} is transferred as c_{ri} when $v_{ri} = 0$. We next turn to such mutation probabilities.

6.2.1 MUTATION MODELS

The frequency with which mutations happen varies hugely between different types of markers. Data for 18 markers available in STRbase[4] show mutation rates from 0.01% to 0.64%. SNP markers tend to have mutation rates in the range 10^{-9} to 10^{-8}. If only a few such markers are used, it may be reasonable to assume absence of mutations. However, the STR markers popular in family genetics cases may have mutation rates around 0.005. This means that mutations occur in a nonignorable proportion of standard cases.

When mutations occur, they may dramatically influence the results, as we saw in Section 2.4. In particular, this happens when a mutation results in an apparent exclusion; see Table 2.5. There is a large difference between assuming that something is impossible (e.g., assigning the probability zero to it) and assigning some small positive probability to it. We also saw in Section 2.4 how the value of this small probability crucially influences the final likelihood results. This underlines the importance of these small probabilities, and they are the subject of this section.

In practical family-genetic work, it may be tempting to ignore mutations in all cases where they do not seem to occur, and consider them only when one or more of the pedigrees considered are excluded by the data. However, that there is no exclusion does not mean that a mutation has not occurred. A mutation may easily lead to genotypes that could be explained even without assuming there are mutations. Such mutations are called *hidden mutations*. Ignoring them may lead to biased results, and a detailed analysis is presented in [102]. In particular, mutation rates are generally underestimated with conventional count-based approaches. In Section 2.4.4 we recommended that one should always use pedigree-level models that include the possibility of mutations, at least when one is using STR markers, even in cases with no apparent exclusions.

Looking again at the example summarized in Table 2.5, we realize that the results depend crucially on the probability that allele 14 or allele 15 mutates to allele 16 or allele 17 during a transmission from father to child. Even if mutations in general happen fairly often, such a specific event is quite rare, even for STR markers. It is quite close to impossible to gather enough data for each specific mutation to be observed enough times to reliably estimate the probability from the frequency. In particular long mutations like 10 to 18 are very rare. Thus, we must rely on combining observations with models which encode what we believe to be true about mutations.

To describe mutation probabilities, we use the following notation: We assume there are n possible alleles at the marker, and that a mutation changes the allele from type i to type j. Specifically, we let m_{ij} be the probability that allele i is transmitted to the child as allele j. Thus, m_{ii} is the probability that allele i will *not* mutate when transferred to the child. If we use R to denote the *mutation rate* and assume that the probability of mutation is independent of which allele we start with, we get

[4] http://www.cstl.nist.gov/strbase/mutation.htm.

$m_{ii} = 1 - R$. The values of m_{ij} for $i \neq j$ specify the particular mutation probabilities mentioned above. In fact, we can collect all these values into the *mutation matrix*

$$\begin{bmatrix} m_{11} & m_{12} & m_{13} & \cdots & m_{1n} \\ m_{21} & m_{22} & m_{23} & \cdots & m_{2n} \\ m_{31} & m_{32} & m_{33} & \cdots & m_{3n} \\ \vdots & \vdots & \vdots & & \vdots \\ m_{n1} & m_{n2} & m_{n3} & \cdots & m_{nn} \end{bmatrix},$$

which specifies the mutation model. Note that each row represents the probabilities of all the possibilities for what happens to allele i, so these values must be nonnegative and sum to 1: for each i, we have

$$m_{i1} + m_{i2} + \cdots + m_{in} = 1.$$

How can we find a mutation matrix that represents the probabilities of various specific mutations for a locus? As noted above, it is very difficult if not impossible to estimate the probability of each specific mutation directly as an observed frequency. Instead, we need to use *mutation models*, where knowledge of how mutations appear to happen is used to make a parameterized model of the mutation matrix, with only a few parameters.

The "Equal" model

Given that a mutation has occurred, what is the probability that it has mutated to each of the other possible alleles at the locus? If we have no particular information about the alleles, it seems simplest to assign equal probability to each possibility. If we also believe that all alleles have equal probability for a mutation to happen in the first place, we get what may be called the "Equal" mutation model. Its mutation matrix is

$$\begin{bmatrix} 1-R & \frac{R}{n-1} & \frac{R}{n-1} & \cdots & \frac{R}{n-1} \\ \frac{R}{n-1} & 1-R & \frac{R}{n-1} & \cdots & \frac{R}{n-1} \\ \frac{R}{n-1} & \frac{R}{n-1} & R-1 & \cdots & \frac{R}{n-1} \\ \vdots & \vdots & \vdots & & \vdots \\ \frac{R}{n-1} & \frac{R}{n-1} & \frac{R}{n-1} & \cdots & 1-R \end{bmatrix},$$

where R is the mutation rate and n is the total number of possible alleles. The model can also be expressed as

$$m_{ij} = \begin{cases} 1-R & \text{if } i = j, \\ \frac{R}{n-1} & \text{if } i \neq j. \end{cases}$$

The "Equal" model is well suited to situations where we have no information that seems to indicate that one type of mutation is likelier than another. For example, this model is reasonable for SNP markers. However, for STR markers, we have a good idea of how mutations happen, and we know that some mutation types are likelier than others. For such markers, we should instead consider the following model.

The "Stepwise" model

STR alleles consist of a number of repetitions of a short sequence of nucleotides; see the glossary and http://www.cstl.nist.gov/strbase/ for more detailed information. When they are copied in the meiosis transferring the allele from parent to child, a kind of "slippage" may occur. Such a copying error most often consists in one sequence being either added or removed. More rarely, two or more sequences are added or removed. We would like the mutation model to reflect these differences.

A simple way to build a model is to say that an addition or subtraction of $k+1$ sequences is r times as probable as an addition or subtraction of k sequences, for any $k>0$, where $r<1$ is some parameter (called "mutation range" in `Familias`[5]). If we assume that our alleles are named with use of the number of sequences in the allele, and then ordered so that the names represent a sequence of consecutive integers, we get the following mutation matrix:

$$\begin{bmatrix} 1-R & k_1 r^1 & k_1 r^2 & \dots & k_1 r^{n-1} \\ k_2 r^1 & 1-R & k_2 r^1 & \dots & k_2 r^{n-2} \\ k_3 r^2 & k_3 r^1 & 1-R & \dots & k_3 r^{n-3} \\ \vdots & \vdots & \vdots & & \vdots \\ k_n r^{n-1} & k_n r^{n-2} & k_n r^{n-3} & \dots & 1-R \end{bmatrix}.$$

Here, R is the mutation rate and $k_1, k_2, \dots, k_n$ are constants chosen so that each line sums to 1. We call this the "Stepwise" model; it can also be expressed as

$$m_{ij} = \begin{cases} 1-R & \text{if } i=j, \\ k_i r^{|i-j|} & \text{if } i \neq j. \end{cases}$$

STR markers may also have *microvariants*. These are alleles that, in addition to a number of repetitions of the STR sequence, also contain some fraction of the sequence. For example, we may have a number of repetitions of the sequence AATC, and in addition the single nucleotide A somewhere in between the repetitions. If, for example, there are 20 repetitions of AATC, the allele with the extra nucleotide would be named 20.1. If there had been two extra nucleotides, we would have used the notation 20.2.

An allele containing an integer number of sequences may mutate to a microvariant, but such mutations are much rarer than mutations to other alleles with an integer number of sequences. We would like our mutation model to reflect this knowledge. Specifically, we subdivide the alleles of the locus into microvariant groups, where each group consists of alleles that differ by an integer number of repetitions of sequences. For example, if we believe the marker TH01 has the alleles

$$5, 6, 7, 8, 8.3, 9, 9.3, 10, 10.3, 11$$

it would have the two groups 5, 6, 7, 8, 9, 10, 11 and 8.3, 9.3, 10.3. Mutations between groups are much rarer than mutations within groups.

[5]Starting from this chapter, `Familias` by default refers to the `R` version.

Below is an illustration of what a mutation matrix with the *extended stepwise model* might look like for a hypothetical marker. The first three columns/rows and the last column and row represent markers in one microvariant group, while the middle three columns/rows represent three consecutive markers in another microvariant group.

$$\begin{bmatrix} 1-R-R_2 & k_1 r^1 & k_1 r^2 & \dots & \frac{R_2}{s_1} & \frac{R_2}{s_1} & \frac{R_2}{s_1} & \dots & k_1 r^{n-1} \\ k_2 r^1 & 1-R-R_2 & k_2 r^1 & \dots & \frac{R_2}{s_2} & \frac{R_2}{s_2} & \frac{R_2}{s_2} & \dots & k_2 r^{n-2} \\ k_3 r^2 & k_3 r^1 & 1-R-R_2 & \dots & \frac{R_2}{s_3} & \frac{R_2}{s_3} & \frac{R_2}{s_3} & \dots & k_3 r^{n-3} \\ \vdots & \vdots & \vdots & & \vdots & \vdots & \vdots & & \vdots \\ \frac{R_2}{s_{i-1}} & \frac{R_2}{s_{i-1}} & \frac{R_2}{s_{i-1}} & \dots & 1-R-R_2 & k_{i-1} r^1 & k_{i-1} r^2 & \dots & \frac{R_2}{s_{i-1}} \\ \frac{R_2}{s_i} & \frac{R_2}{s_{i1}} & \frac{R_2}{s_i} & \dots & k_i r^1 & 1-R-R_2 & k_i r^1 & \dots & \frac{R_2}{s_i} \\ \frac{R_2}{s_{i+1}} & \frac{R_2}{s_{i+1}} & \frac{R_2}{s_{i+1}} & \dots & k_{i+1} r^2 & k_{i+1} r^1 & 1-R-R_2 & \dots & \frac{R_2}{s_{i+1}} \\ \vdots & \vdots & \vdots & & \vdots & \vdots & \vdots & & \vdots \\ k_n r^{n-1} & k_n r^{n-2} & k_n r^{n-3} & \dots & \frac{R_2}{s_n} & \frac{R_2}{s_n} & \frac{R_2}{s_n} & \dots & 1-R-R_2 \end{bmatrix}.$$

In the matrix, R is the rate of integer-length mutations, r is the mutation range, R_2 is the rate of fractional-length mutations, s_i is the number of alleles outside allele i's microvariant group for each i, and $k_1, \dots, k_n$ are constants chosen so that each line sums to 1. The entries of the mutation matrix can also be expressed as

$$m_{ij} = \begin{cases} 1-R-R_2 & \text{if } i = j, \\ k_i r^{|i-j|} & \text{if } i \neq j \text{ and } i \text{ and } j \text{ are in the same microvariant group,} \\ \frac{R_2}{s_i} & \text{if } i \neq j \text{ and } i \text{ and } j \text{ are in different microvariant groups.} \end{cases}$$

The model above can be used if all microvariant groups have more than one member. The probability for any allele to mutate will then be $R + R_2$, so this is the total mutation rate. However, if a microvariant group contains only one allele, the mutation rate from that allele will be R_2, which with realistic values for R and R_2 will be a much smaller number than $R + R_2$. On the other hand, the actual probability for such an allele to mutate an integer number of steps, and then to an allele not seen before in the underlying database, is probably on the order of R. So, an even more realistic model should include microvariants other than the one that has been observed, but with very low population frequencies. Use of nonzero frequencies for unobserved alleles was discussed in Section 6.1.1.

Example 6.4 Mutation models. Consider the marker TH01 with alleles as listed above, and consider a duo paternity case where the alleged father has genotype 9.3/9.3, while the daughter has genotype 5/5. Below we find the LR in favor of paternity for a mutation rate of 0.005 and the "Equal" model, or integer and fractional mutation rates of 0.005 and 0.0000001, respectively, and the "Stepwise" model, with the default mutation range parameter 0.5:

```
data(NorwegianFrequencies)
loc1 <- FamiliasLocus(NorwegianFrequencies$TH01,
        MutationRate = 0.005, MutationModel = "Equal")
loc2 <- FamiliasLocus(NorwegianFrequencies$TH01,
        MutationRate = 0.005, MutationRate2 =0.0000001,
        MutationModel = "Stepwise")
ped1 <- FamiliasPedigree(c("CH", "AF"), c(NA, NA), c(NA, NA),
        c("female", "male"))
ped2 <- FamiliasPedigree(c("CH", "AF"), c("AF", NA), c(NA,
        NA), c("female", "male"))
peds <- list(notFather = ped1, isFather = ped2)
data <- rbind(CH = c(5,5), AF = c(9.3, 9.3))
```

This results in the output

```
FamiliasPosterior(peds, loc1, data)$LR
```

```
notFather  isFather
1.0000000 0.1836397
```

```
FamiliasPosterior(peds, loc2, data)$LR
```

```
   notFather    isFather
1.000000e+00 4.722163e-06
```

We see that there is a huge difference in the results from the `FamiliasPosterior` function between the two mutation models. For the “Equal” model, the LR may be unexpectedly large: Even if there is an obvious inconsistency between the alleged father and the child, the LR is as large as 0.18. The reason is that allele 5 has a very low probability, as seen in the list

```
NorwegianFrequencies$TH01
```

```
           5            6            7            8          8.3
0.0030252481 0.2146878655 0.2080638697 0.0922696422 0.0001564779
           9          9.3           10         10.3           11
0.1333719165 0.3404967867 0.0077717151 0.0001043189 0.0000521594
```

whereas allele 9 has a quite high probability. So allele 5 is almost as well explained as coming from a mutation of 9.3 as coming from the population directly. On the other hand, this depends on a mutation from 9.3 to 5 being plausible. In the “Equal” mutation model, the probability of one happening is $0.005/(10 - 1) = 0.00056$. According to the “Stepwise” model, however, it is not, as the mutation changes the microvariant group. In fact, the probability is now $R_2/7 = 0.0000001/7 = 1.428571 \times 10^{-8}$.

Note how (as we saw in Section 2.4.3) we can compute the LR directly in this simple case, without the `FamiliasPosterior` function. Indeed Equation 2.12 simplifies to

$$\text{LR} = \frac{1}{4}\frac{(m_{ac} + m_{bc})p_d + (m_{ad} + m_{bd})p_c}{p_c p_d} = \frac{m_{9.3,5}}{p_5},$$

which is used below:

```
loc1$maleMutationMatrix["9.3", "5"]/NorwegianFrequencies$
          TH01 ["5"]
```

```
        5
0.1836397
```

```
loc2$maleMutationMatrix["9.3", "5"]/NorwegianFrequencies$
          TH01 ["5"]
```

```
           5
4.722163e-06
```

*Stationary mutation models

In the introduction to this chapter, we noted that the subdivision of our model for genotypes of tested persons into a pedigree-level model and a population-level model is rather arbitrary. For example, if to a pedigree we add a father and a mother to one of the founders, we should ideally obtain unchanged probability for observing specific genotypes of tested persons. However, because our population-level model and pedigree-level model may not be completely compatible, this is not necessarily so. Specifically, the mutation model is usually not compatible with the population-level model, in the sense that we may get different answers above.

Theoretically, the sequence of transfers of an allele through the generations can be viewed as a Markov chain, as the type of the allele depends on the type of the allele in the previous generation, via the mutation model, and not on the allele in generations before that given the allele in the previous generation. Under some very general assumptions, we know[6] that the distribution for the type of the allele will converge to a unique *stationary distribution*. In other words, no matter what allele we start with, if we let it mutate using our mutation model through enough generations, we will obtain a certain probability distribution of alleles. If this stationary distribution is equal to the population distribution for the allele, adding extra parents to pedigrees will not change the likelihood results for pedigree computations. But if the stationary distribution differs from the population distribution, there will be slight differences in the results.

Such differences, although generally small, can be an inconvenience, as the authority of a method is undermined when it gives different results in situations where people believe it should give identical results. For this reason, one may consider using *stationary* mutation models, by which we mean mutation models whose stationary distribution is equal to the population distribution of frequencies. Below, we consider some options for obtaining such models.

*Model based on frequencies

Let $M = \{m_{ij}\}$ be the mutation matrix, and let v be a (row) vector signifying probabilities for the n alleles. Then the probability distribution for the alleles after

[6]See, for example, http://en.wikipedia.org/wiki/Stochastic_matrix.

one generation is given by the matrix product vM, and after k generations it will be vM^k. If p is the vector of population probabilities for the alleles, a mutation model is stationary if and only if $pM = p$.

Let $\mathbf{1} = (1, \ldots, 1)$ be the row vector of n 1's. We can specify a mutation model as

$$M_{\mathrm{f}} = (1 - c)I + c\mathbf{1}^t p, \tag{6.12}$$

where $c = R/(1 - pp^t) = R/(1 - \sum_{i=1}^n p_i^2)$. This mutation matrix is then stationary, and has *expected mutation rate* R (see Exercise 6.1). By "expected mutation rate" we mean the probability of a mutation given that the state of an allele is given by the probability distribution p. We may also specify the mutation matrix as

$$m_{ij} = \begin{cases} 1 - \frac{R(1-p_i)}{1-\sum_{i=1}^n p_i^2} & \text{if } i = j, \\ \frac{Rp_j}{1-\sum_{i=1}^n p_i^2} & \text{if } i \neq j. \end{cases}$$

The mutation model above has the name "Proportional" in `Familias`.

*Stabilizing existing mutation models

In the above mutation model, the relative probabilities of mutations of various alleles, given that a mutation happens, depends only on their relative population frequencies. That may not be optimal; we have discussed how mutation probabilities for STR markers in reality depend strongly on how many sequences are added or removed in the mutation. So a different idea is to take an existing mutation model and "stabilize" it—that is, make it stationary, but otherwise change it as little as possible.

If M is a mutation matrix, define the matrix

$$M^* = \frac{1 - \mathrm{Tr}(D(p)M)}{1 - \mathrm{Tr}(D(v)M)} D(p)^{-1} D(v)(M - I) + I,$$

where we write $D(x)$ for the $n \times n$ diagonal matrix with the vector x along its diagonal, I for the identity matrix, Tr for the trace operator, p for the vector of population frequencies, and v for the stationary distribution of M. One may show that, as long as all its elements are positive, M^* is a stationary mutation matrix with the same expected mutation rate as M. One may also see that the ratio between off-diagonal elements in the same row is unchanged from M to M^*.

A problem with the stabilization method above is that diagonal elements in the stabilized mutation matrix—that is, the elements indicating the probability of no mutation—may become quite small, and sometimes negative. This problem occurs quite easily, for example when the method is applied to the data in `NorwegianFrequencies`, and is an obvious argument against use of the stabilization above.

6.3 OBSERVATIONAL-LEVEL MODELS

We now turn our attention to models for $\Pr(\mathcal{D} \mid \mathcal{G})$, the probability of the observed data $\mathcal{D}$ given the phased genotypes $\mathcal{G}$ of the DNA-tested persons. In many applications, the available data give fairly certain knowledge of the *unphased genotypes* of the tested persons, so $\Pr(\mathcal{D} \mid \mathcal{G})$ models only the relationship between phased genotypes (genotypes where it is known which alleles are maternal and which alleles are paternal) and unphased genotypes (where this information is not available). But in some applications, one may have to consider the possibility of various types of observational errors. This can be due to, for example, poor quality of the DNA used for testing, or this DNA being a mixture from different sources.

We may subdivide observational-level models into two types: summation models and probabilistic models. If $\Pr(\mathcal{D} \mid \mathcal{G})$ can take as values only 0 or 1, we call the model a summation model. In the latter case, if $G_1, G_2, \ldots, G_J$ are the values of $\mathcal{G}$ that make $\Pr(\mathcal{D} \mid \mathcal{G}) = 1$ for a specific data value $\mathcal{D}$, we get from Equation 6.1

$$\begin{aligned}\Pr(\mathcal{D} \mid \mathcal{P}) &= \sum_{\mathcal{G}} \sum_{\mathcal{F}} \Pr(\mathcal{D} \mid \mathcal{G}) \Pr(\mathcal{G} \mid \mathcal{F}, \mathcal{P}) \Pr(\mathcal{F}) \\ &= \sum_{j=1}^{J} \sum_{\mathcal{F}} \Pr(G_j \mid \mathcal{F}, \mathcal{P}) \Pr(\mathcal{F}). \end{aligned} \tag{6.13}$$

Example 6.5 Unphased genotype data. In most cases of pedigree inference, one has available reliable data about the unphased genotypes of each tested person. Computations of likelihoods can be done by summing over the likelihoods of all possible compatible phased genotypes. Generally, one needs to sum over all possible combinations of haplotypes G_i for the tested persons, and there can be a large number of such haplotypes if many markers are used. Assume n persons are tested at m autosomal markers, and for simplicity assume that all these persons are heterozygote in all m markers. Then there are a total of $J = (2^m)^n = 2^{mn}$ possible phased genotypes to sum over. However, if markers can be treated independently, there would be only 2^n phased genotypes to sum over at each marker, making computations much easier.

Example 6.6 Standard duo case with possibility of mutations. Assume we have a standard duo case, with data for a single marker. The alleged father has observed genotype a/b and the child has observed genotype c/d (where a, b, c, and d need not be different). We write p_a, p_b, etc., for the fixed known population frequencies, and, for example, m_{ac} for the probability that an allele a is transferred (i.e., possibly mutated) to an allele c. If $\mathcal{P}$ is the pedigree of paternity and we write, for example, (a, b) for the phased genotype with paternal allele a and maternal allele b, we get

$$\begin{aligned}\Pr(\mathcal{D} \mid \mathcal{P}) &= \Pr(\text{AF} = a/b, \text{CH} = c/d \mid \mathcal{P}) \\ &= \Pr(\text{AF} = (a, b), \text{CH} = (c, d) \mid \mathcal{P}) \end{aligned} \tag{6.14}$$

$$
\begin{aligned}
&+ I(a \neq b) \Pr(\mathrm{AF} = (b,a), \mathrm{CH} = (c,d) \mid \mathcal{P}) \\
&+ I(c \neq d) \Pr(\mathrm{AF} = (a,b), \mathrm{CH} = (d,c) \mid \mathcal{P}) \\
&+ I(a \neq b) I(c \neq d) \Pr(\mathrm{AF} = (b,a), \mathrm{CH} = (d,c) \mid \mathcal{P}) \\
= {} & p_a p_b p_d \frac{1}{2}(m_{ac} + m_{bc}) \\
&+ I(a \neq b) p_a p_b p_d \frac{1}{2}(m_{ac} + m_{bc}) \\
&+ I(c \neq d) p_a p_b p_c \frac{1}{2}(m_{ad} + m_{bd}) \\
&+ I(a \neq b) I(c \neq d) p_a p_b p_c \frac{1}{2}(m_{ad} + m_{bd}) \\
= {} & 2^{-I(a=b)-I(c=d)} p_a p_b \left(p_d(m_{ac} + m_{bc}) + p_c(m_{ad} + m_{bd})\right).
\end{aligned}
$$

Example 6.7 Silent alleles. Silent alleles were introduced in Section 2.6 as alleles that cannot be detected with a specific technique or DNA testing kit. In practice, however, an electrophoresis diagram may help distinguish between data from a homozygote with genotype a/a, where a is some non-silent allele, and a heterozygote with genotype a/S, where S is the silent allele, as the response from the former would be expected to be twice as strong (i.e., produce twice as high a peak) as the response from the latter. If one suspects that there is a silent allele in the case data, one may also proceed to analyze the data with a different method or kit, to be able to detect the silent allele.

Below, we will look at the simplified model where a homozygote with genotype a/a and a heterozygote with genotype a/S generate indistinguishable data denoted $a/-$. This is the model implemented in the `Familias` programs. Clearly, if the observed data correspond to heterozygote a/b with $a \neq b$, the true genotype must be a/b. A person homozygous in the silent allele would not appear to have a genotype at all. However, as silent alleles generally have low population frequencies, such persons are very rare. In terms of computations of likelihoods, $\mathcal{G}$ would consist of the three possible phased genotypes (a,a), (a,S), and (S,a) for every observed $a/-$. Likelihoods can be computed as sums of the likelihoods where each observed value $a/-$ is replaced with a true homozygote of type a/a and a heterozygote of type a/S.

Example 6.8 Combining silent alleles and mutations in a duo case. We build on Example 6.6, assuming that there is now also a silent allele S, with population frequency p_S. We assume that mutation rates to and from allele S are zero. We now get

$$
\begin{aligned}
\Pr(\mathcal{D} \mid \mathcal{P}) &= \Pr(\text{AF observed as } a/b, \text{CH observed as } c/d \mid \mathcal{P}) \\
&= \Pr(\mathrm{AF} = a/b, \mathrm{CH} = c/d \mid \mathcal{P}) \\
&\quad + I(a = b) \Pr(\mathrm{AF} = a/S, \mathrm{CH} = c/d \mid \mathcal{P}) \\
&\quad + I(c = d) \Pr(\mathrm{AF} = a/b, \mathrm{CH} = c/S \mid \mathcal{P}) \\
&\quad + I(a = b) I(c = d) \Pr(\mathrm{AF} = a/S, \mathrm{CH} = c/S \mid \mathcal{P})
\end{aligned}
$$

$$
\begin{aligned}
&= 2^{-I(a=b)-I(c=d)} p_a p_b \left(p_d(m_{ac}+m_{bc})+p_c(m_{ad}+m_{bd})\right) \\
&\quad + I(a=b)2^{-I(c=d)} p_a p_S \left(p_d m_{ac}+p_c m_{ad}\right) \\
&\quad + I(c=d)2^{-I(a=b)} p_a p_b p_S (m_{ac}+m_{bc}) \\
&\quad + I(a=b)I(c=d) p_a p_S (p_S m_{ac}+p_c).
\end{aligned}
$$

The last expression can be simplified in various ways.

Example 6.9 Dropouts. When the genetic test is based on small amounts of DNA, or DNA of low quality, there may be a nonnegligible probability of errors in the observation of unphased genotypes. An allele present in the genotype may drop out—that is, not be observed at all—or there may even be cases of drop-ins, where alleles not present in the genotype are nonetheless observed, owing to problems in the polymerase chain reaction process (e.g., stutters) or DNA contamination.

A simple model for dropouts is as follows (see also Section 3.3.2): Each allele has a fixed probability d for not being observed, and whether or not an allele is observed is independent between alleles. As with silent alleles, if the observed genotype is heterozygous a/b, the true genotype must be heterozygous a/b. However, the observed genotype $a/-$ is compatible with the three phased genotypes (a,a), (a,b), and (b,a), where $b \neq a$, and

$$
\begin{aligned}
\Pr(a/- \mid (a,a)) &= 1-d^2, \\
\Pr(a/- \mid (a,b)) &= (1-d)d, \\
\Pr(a/- \mid (b,a)) &= d(1-d).
\end{aligned}
$$

Example 6.10 Genotyping errors. Whereas STR markers are usually very reliably typed, there tends to be nonignorable genotyping error rates for SNP markers. The simplest way to model this, for diallelic markers, is that each SNP allele has a fixed probability e of being mistyped to the alternative allele, independent of other alleles. See also Example 3.3. If the data at a single SNP marker correspond to heterozygote a/b, we have

$$
\begin{aligned}
&\Pr(\text{observed } a/b \mid (a,a)) = 2(1-e)e, \\
&\Pr(\text{observed } a/b \mid (a,b)) = \Pr(\text{observed } a/b \mid (b,a)) = (1-e)^2+e^2, \\
&\Pr(\text{observed } a/b \mid (b,b)) = 2e(1-e),
\end{aligned}
$$

whereas if the data correspond to a homozygote of type a/a, we have

$$
\begin{aligned}
&\Pr(\text{observed } a/a \mid (a,a)) = (1-e)^2, \\
&\Pr(\text{observed } a/a \mid (a,b)) = \Pr(\text{observed } a/a \mid (b,a)) = (1-e)e, \\
&\Pr(\text{observed } a/a \mid (b,b)) = e^2.
\end{aligned}
$$

Example 6.11 Summation models for mixtures. In most applications of relationship inference, separate data are available for each tested person. However, in some cases containing both identification and relationship inference, some data may be based on the genotypes of several persons; we say we have a mixture. Under some hypotheses, one or more of these genotypes may be from otherwise unknown and untested persons, and calculations may then become quite complex.

DNA mixtures are often of a type which necessitates modeling dropouts or other observational errors. It may also be possible to discern which person has contributed which alleles when different persons have contributed different amounts of DNA to the mix. The size of a response for a mix then contains information about who contributed the allele. But if we assume we can ignore such complications, mixtures can be modeled with a summation model: For every marker, a list of alleles has been observed in the data. In our computations, we sum the likelihoods of (phased) genotypes of contributing persons that jointly contain exactly the observed alleles. For example, if alleles 15, 19, and 21 are observed in a mix and the hypothesis is that two persons have contributed, there are $3 \times 4 \times 3 = 36$ possible compatible phased genotypes: there are three alleles that may be duplicated, and for each such duplication the two other alleles have 3×4 positions where they may be placed in the phased genotypes.

Many more advanced observational-level models exist, in particular for model mixtures. Commercial programs implement probabilistic models taking into account electrophoresis top heights in the data. Although most implementations are directed toward pure identification problems, the models can clearly also be used in connection with relationship inference. We will, however, not discuss such models further in this book.

6.4 COMPUTATIONS

In this chapter, we have presented a number of alternatives for population-level, pedigree-level, and observational-level models. Combining choices for such models, Equation 6.1 defines the likelihood we are interested in computing. The choice of models—that is, which restrictive assumptions it is reasonable to make—will, of course, depend on the context. The decision should be based on what complications could realistically be expected, and to what extent such complications may affect the results. One should not ignore a possible complication just because it is not apparent in the data. For example, mutations are quite common for STR markers, so models implementing mutations should be used even if the case data do not seem to indicate, for example, an exclusion which may be caused by a mutation. Use of models for mutations only when the case data indicate exclusions creates a bias; see, for example, [102]. Similarly, when low-quality test DNA or small amounts of DNA make the possibility of dropouts realistic, models implementing dropouts should be used generally, and not only in cases where inspection of the data leads to a suspicion of dropouts.

Once a model has been chosen, Equation 6.1 is rarely a realistic way to perform the actual computations, as the number of terms is most often too large. Instead, smarter solutions employing various types of tricks are used. Which tricks are possible depends strongly on the model chosen. In particular, two choices are important: whether one assumes independence between markers, and whether one allows the possibility of mutations within the pedigree. Assuming there is independence between markers, both in the population and within the pedigree, allows the likelihood to be computed as a product over the likelihoods for each marker, thereby hugely simplifying computations. Assuming there are no mutations allows the use of IBD ideas, which may lead to large computational simplifications.

Example 6.12 Using unphased genotypes. Standard methods of observing genotypes cannot distinguish between paternal and maternal alleles or haplotypes. Mendelian inheritance also does not distinguish between these two. These facts together mean that all our population-level, pedigree-level, and observational-level models may be formulated in terms of ordinary unphased genotypes. Use of unphased genotypes may mean that one needs to sum over slightly fewer alternative genotypes. However, model formulations generally becomes slightly trickier.

Often, computational tricks can be formulated as a way to use conditional independencies. The Elston-Stewart algorithm, first mentioned in Section 2.3, may be seen as an example of this. Let $\mathcal{G}_1$ and $\mathcal{G}_2$ be the genotypes of two separate sets of persons (or alleles or haplotypes) in the pedigree. If there is a set of persons such that for their genotypes $\mathcal{G}_3$ we have

$$\Pr(\mathcal{G}_2 \mid \mathcal{G}_1, \mathcal{G}_3) = \Pr(\mathcal{G}_2 \mid \mathcal{G}_3),$$

then that set of persons can be called a cutset. If, for example, one has a cutset separating the DNA-tested persons from the founders, we may compute, writing $\mathcal{C}$ for the genotype of the cutset,

$$\Pr(\mathcal{G} \mid \mathcal{F}) = \sum_{\mathcal{C}} \Pr(\mathcal{G} \mid \mathcal{C}) \Pr(\mathcal{C} \mid \mathcal{F}).$$

If the number of possible values for $\mathcal{C}$ is limited, the equation above may be a very valuable computational shortcut.

In a similar way, the Lander-Green algorithm uses conditional independencies along the sequence of markers to facilitate computations.

In the rest of this section, we will look at another very important idea for computational simplifications: under certain conditions, the genotypes of DNA-tested persons are conditionally independent of the pedigree given the information on how probable it is that they have inherited the same alleles within the pedigree.

6.4.1 IDENTICAL BY DESCENT

The concept of IBD was introduced in Section 2.3.3. A more general introduction is given below. Recall from Section 6.2 the definition of the variables $V_1, \ldots, V_R$: for each of the $r = 1, \ldots, R$ parent-child relationships in the pedigree, V_r describes whether, for each locus, the paternal or the maternal allele has been transmitted. Similarly to Equation 6.10 we get

$$\Pr(\mathcal{G} \mid \mathcal{P}) = \sum_{V_1,\ldots,V_R} \Pr(\mathcal{G} \mid V_1,\ldots,V_R) \prod_{r=1}^{R} \Pr(V_r). \tag{6.15}$$

In other words, we may compute the probability of the phased data $\mathcal{G}$ given the pedigree by first computing, for all possible combinations of the values of $V_1,\ldots,V_R$, the probability of the genotype given this combination, before multiplying it by the probabilities $\Pr(V_r)$ and then summing over all possibilities for $V_1,\ldots,V_R$. If we make the assumption that there are no mutations within the pedigree, then

$$\Pr(\mathcal{G} \mid V_1,\ldots,V_R) = \sum_{\mathcal{F}} \Pr(\mathcal{G} \mid V_1,\ldots,V_R,\mathcal{F}) \Pr(\mathcal{F})$$

can be computed fairly directly: We first check whether alleles in $\mathcal{G}$ that have been inherited from the same founder allele according to $V_1,\ldots,V_R$ are in fact identical. If they are not, $\Pr(\mathcal{G} \mid V_1,\ldots,V_R) = 0$. Otherwise, $\Pr(\mathcal{G} \mid V_1,\ldots,V_R,\mathcal{F})$ is 0 or 1 depending on whether the founder alleles in $\mathcal{F} = (f_1,\ldots,f_F)$ that are transferred by $V_1,\ldots,V_R$ match those in $\mathcal{G}$. Thus $\Pr(\mathcal{G} \mid V_1,\ldots,V_R)$ is a sum over terms $\Pr(\mathcal{F})$, with one term for each $\mathcal{F}$ containing the founder alleles compatible with $\mathcal{G}$.

We can improve on this computation in the following way: Consider the sequence of the $2J$ alleles of the phased genotypes at one specific marker of the J tested persons. Each choice of $V_1,\ldots,V_R$ partitions[7] this sequence into subsets of alleles inherited from the same founder allele. This partition for each marker, together with the index of the founder allele each subset is inherited from in the sequence of F founder alleles, is enough to compute $\Pr(\mathcal{G} \mid V_1,\ldots,V_R,\mathcal{F})$. We now make the assumption that our population-level model assumes independence between markers, and that the probability of a sequence of founder alleles at one marker is unchanged under permutations of the order of the sequence. Then we do not need to keep track of the index of the founder allele each allele subset is inherited from in order to compute $\Pr(\mathcal{G} \mid V_1,\ldots,V_R)$.

Let K be a variable that for each choice $V_1,\ldots,V_R$ records how the sequence of alleles in $\mathcal{G}$ is partitioned into subsets for each marker. We may split K into marker-specific variables $K = (K_1,\ldots,K_m)$, where the possible values for each K_i are the possible partitions of the sequence $1,2,\ldots,2J$. The variable K now carries the information necessary to compute the probability of the phased genotype $\mathcal{G}$: instead of Equation 6.15 we may write

$$\Pr(\mathcal{G} \mid \mathcal{P}) = \sum_{K} \Pr(\mathcal{G} \mid K) \Pr(K \mid \mathcal{P}), \tag{6.16}$$

[7]A partition of a set is a subdivision of the set into non-overlapping subsets covering the whole set.

which is a computational improvement, as there are fewer possible values for K than for the sequence $V_1, \ldots, V_R$. Indeed, we may further improve this: Derive the variable $\overline{K}$ from K by identifying all partitions that are equal after switching some of the pairs of paternal and maternal alleles. Let $\overline{\mathcal{G}}$ be the unphased genotype corresponding to $\mathcal{G}$. Then the probability of $\overline{\mathcal{G}}$ will depend only on the value of $\overline{K}$, and we can write

$$\Pr(\overline{\mathcal{G}} \mid \mathcal{P}) = \sum_{\overline{K}} \Pr(\overline{\mathcal{G}} \mid \overline{K}) \Pr(\overline{K} \mid \mathcal{P}), \tag{6.17}$$

which is an improvement over Equation 6.16 in the sense that $\overline{K}$ has fewer possible values than K.

Below, we consider several concrete examples of the ideas above.

Assuming independent markers, two tested persons, and no inbreeding

When we assume independence between markers, we may use Equation 6.17 for each marker and then multiply the results. The possible values for K are then all possible partitions of the sequence $1, 2, \ldots, 2J$. When we assume that there are $J = 2$ tested persons, we consider all possible partitions of the sequence $1, 2, 3, 4$. One may check that there are 15 such *Jacquard* partitions:

$$
\begin{aligned}
&\{\{1,2,3,4\}\},\\
&\{\{1,2\},\{3,4\}\},\\
&\{\{1,2,3\},4\},\{\{1,2,4\},3\},\\
&\{\{1,2\},3,4\},\\
&\{\{1,3,4\},2\},\{\{2,3,4\},1\},\\
&\{1,2,\{3,4\}\},\\
&\{\{1,3\},\{2,4\}\},\{\{1,4\},\{2,3\}\},\\
&\{\{1,3\},2,4\},\{\{1,4\},2,3\},\{\{2,3\},1,4\},\{1,3,\{2,4\}\},\\
&\{1,2,3,4\}.
\end{aligned}
$$

For instance, $\{\{1,2\},\{3,4\}\}$ denotes that alleles 1 and 2 in individual 1 are IBD and that alleles 3 and 4 in individual 2 are IBD as illustrated in column 2 of Figure 6.5, which shows the *condensed* Jacquard partitions discussed below.

Given a pedigree, we say that a person in it is inbred if his or her parents have a common ancestor within the pedigree. In other words, a person is inbred if his or her two alleles may be IBD within the pedigree. If we exclude this possibility, we exclude all possible values of K where either 1 and 2 is in the same partition, or 3 and 4 is in the same partition. We are then left with the partitions in the three last lines of the above list. Let us use overline notation to indicate the equivalence class of a partition when we identify partitions by switching maternal and paternal alleles. Then we get

$$\overline{\{\{1,3\},2,4\}} = \overline{\{\{1,4\},2,3\}} = \overline{\{\{2,3\},1,4\}} = \overline{\{\{1,3\},2,4\}}, \tag{6.18}$$

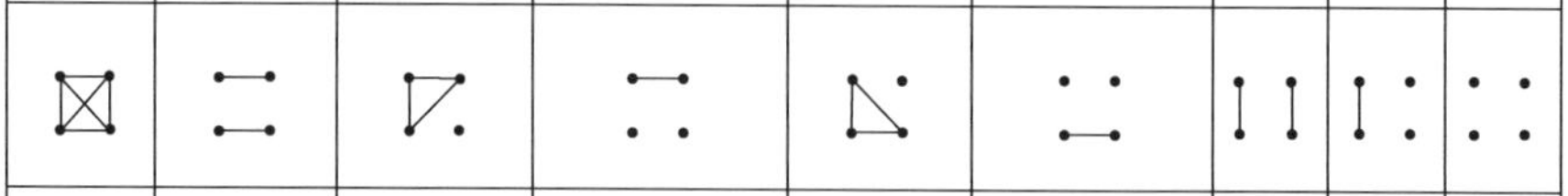

Σ_1	Σ_2	Σ_3	Σ_4	Σ_5	Σ_6	Σ_7	Σ_8	Σ_9
$\begin{bmatrix} p & 0 & 0 \\ 0 & 0 & 0 \\ 0 & 0 & q \end{bmatrix}$	$\begin{bmatrix} p^2 & 0 & pq \\ 0 & 0 & 0 \\ pq & 0 & q^2 \end{bmatrix}$	$\begin{bmatrix} p^2 & pq & 0 \\ 0 & 0 & 0 \\ 0 & pq & q^2 \end{bmatrix}$	$\begin{bmatrix} p^3 & 2p^2q & pq^2 \\ 0 & 0 & 0 \\ p^2q & 2pq^2 & q^3 \end{bmatrix}$	$\begin{bmatrix} p^2 & 0 & 0 \\ pq & 0 & pq \\ 0 & 0 & q^2 \end{bmatrix}$	$\begin{bmatrix} p^3 & 0 & p^2q \\ 2p^2q & 0 & 2pq^2 \\ qp^2 & 0 & q^3 \end{bmatrix}$	MZ	PO	UN

FIGURE 6.5

Jacquard condensed identity states $\Sigma_1, \ldots, \Sigma_9$. In the identity diagrams in the second row, the top two points refer to the unordered alleles of individual 1 and the bottom two points refer to the unordered alleles of individual 2. Connected alleles are IBD. In the bottom row of the diagram, we consider a SNP with allele frequencies *p*, *q*; see Exercise 6.4.

and we are left with three possible values of $\overline{K}$:

$$\sigma_0 = \overline{\{1,2,3,4\}},$$
$$\sigma_1 = \overline{\{\{1,3\},2,4\}},$$
$$\sigma_2 = \overline{\{\{1,3\},\{2,4\}\}}.$$

These three values indicate zero alleles IBD, one allele IBD, and two alleles IBD, respectively. Using the notation $\kappa_i = \Pr(\sigma_i \mid \mathcal{P})$, we may rewrite Equation 6.17 as

$$\Pr(\overline{\mathcal{G}} \mid \mathcal{P}) = \Pr(\overline{\mathcal{G}} \mid \overline{K} = \sigma_0)\kappa_0 + \Pr(\overline{\mathcal{G}} \mid \overline{K} = \sigma_1)\kappa_1 + \Pr(\overline{\mathcal{G}} \mid \overline{K} = \sigma_2)\kappa_2,$$

which is then a re-expression of Equation 2.6.

Allowing inbreeding

For a single marker and two tested persons, we found and listed 15 possible values for K in the previous section. If we allow inbreeding, none of these are excluded for that reason. However, some of them are identical when we identify partitions by allowing switches between paternal and maternal alleles. In addition to the identifications of Equation 6.18, we also have that $\overline{\{\{1,2,3\},4\}} = \overline{\{\{1,2,4\},3\}}$ and $\overline{\{\{1,3,4\},2\}} = \overline{\{\{2,3,4\},1\}}$, so we are left with the following nine possible values of $\overline{K}$ shown in Figure 6.5:

$$\Sigma_1 = \overline{\{\{1,2,3,4\}\}},$$
$$\Sigma_2 = \overline{\{\{1,2\},\{3,4\}\}},$$
$$\Sigma_3 = \overline{\{\{1,2,3\},4\}},$$
$$\Sigma_4 = \overline{\{\{1,2\},3,4\}},$$

$$\Sigma_5 = \overline{\{\{1,3,4\},2\}},$$
$$\Sigma_6 = \overline{\{\{3,4\},1,2\}},$$
$$\Sigma_7 = \overline{\{\{1,3\},\{2,4\}\}},$$
$$\Sigma_8 = \overline{\{\{1,3\},2,4\}},$$
$$\Sigma_9 = \overline{\{1,2,3,4\}}.$$

Using the notation $\Delta_i = \Pr(\Sigma_i \mid \mathcal{P})$, we may rewrite Equation 6.17 as

$$\Pr(\overline{\mathcal{G}} \mid \mathcal{P}) = \sum_{i=1}^{9} \Pr(\overline{\mathcal{G}} \mid \Sigma_i)\Delta_i.$$

The numbers $\Delta_1, \ldots, \Delta_9$ are called the "condensed Jacquard coefficients" and are calculated by the `R` package `identity`. See Exercises 6.4 and 6.5.

Allowing linked markers

Our final example concerns sequences of m markers that are not assumed to be independent in the pedigree-level model. As discussed above, K in Equation 6.16 can then be split into a sequence of marker-specific variables, and thus a similar splitting can be done with $\overline{K}$ in Equation 6.17. If we look at the case with two tested persons P_1 and P_2 and no inbreeding, each $\overline{K}_i$ will have three possible values, so $\overline{K}$ will have 3^m possible values. Similarly, if we include the possibility of inbreeding, $\overline{K}$ will have 9^m possible values.

Let us look in detail at the case of two markers with a recombination probability of ρ, and no inbreeding. Then $\overline{K}$ will have the nine possible values

$$\sigma_{0,0} = \left(\overline{\{1,2,3,4\}}, \overline{\{1,2,3,4\}}\right),$$
$$\sigma_{0,1} = \left(\overline{\{1,2,3,4\}}, \overline{\{\{1,3\},2,4\}}\right),$$
$$\sigma_{0,2} = \left(\overline{\{1,2,3,4\}}, \overline{\{\{1,3\},\{2,4\}\}}\right),$$
$$\sigma_{1,0} = \left(\overline{\{\{1,3\},2,4\}}, \overline{\{1,2,3,4\}}\right),$$
$$\sigma_{1,1} = \left(\overline{\{\{1,3\},2,4\}}, \overline{\{\{1,3\},2,4\}}\right),$$
$$\sigma_{1,2} = \left(\overline{\{\{1,3\},2,4\}}, \overline{\{\{1,3\},\{2,4\}\}}\right),$$
$$\sigma_{2,0} = \left(\overline{\{\{1,3\},\{2,4\}\}}, \overline{\{1,2,3,4\}}\right),$$
$$\sigma_{2,1} = \left(\overline{\{\{1,3\},\{2,4\}\}}, \overline{\{\{1,3\},2,4\}}\right),$$
$$\sigma_{2,2} = \left(\overline{\{\{1,3\},\{2,4\}\}}, \overline{\{\{1,3\},\{2,4\}\}}\right).$$

To use Equation 6.17, we need to be able to compute both $\Pr(\overline{\mathcal{G}} \mid \overline{K})$ and $\Pr(\overline{K} \mid \mathcal{P})$ for each of these values. The first can be easily found by looking at each of the two markers separately, and otherwise computing as in previous IBD computations. The values will be functions of the allele frequencies at the two markers. For any pedigree $\mathcal{P}$, we can write the nine values

$$\Pr(\overline{K} \mid \mathcal{P}) = \Pr(\sigma_{i,j} \mid \mathcal{P}) = k_{i,j} \quad \text{for } i,j = 0,1,2 \tag{6.19}$$

in a symmetric table:

$k_{0,0}$	$k_{0,1}$	$k_{0,2}$
$k_{1,0}$	$k_{1,1}$	$k_{1,2}$
$k_{2,0}$	$k_{2,1}$	$k_{2,2}$

Note that the three rows, as well as the three columns, will sum to $\kappa_0, \kappa_1, \kappa_2$, respectively, and that the numbers in the whole table will sum to 1.

We will consider the table above for three pedigrees, shown in Figure 6.6:

$\mathcal{P}_1$: P_1 and P_2 constitute a grandparent-grandchild pair (as 1 and 6 in Figure 6.6).
$\mathcal{P}_2$: P_1 and P_2 are half siblings (as 6 and 8).
$\mathcal{P}_3$: P_1 and P_2 constitute an uncle-nephew relationship (as 3 and 6).

It is interesting to compare these pedigrees because for all of them we have $(\kappa_0, \kappa_1, \kappa_2) = \left(\frac{1}{2}, \frac{1}{2}, 0\right)$. This means that the likelihood for data from unlinked markers will be equal for all of them, and one cannot separate between them using such data. However, as we see below, likelihoods differ when we consider linked markers.

As the three rows and the three columns of the tables for all the pedigrees sum to $1/2$, $1/2$, and 0, respectively, we see that only the four values in the upper left-hand corner of their tables will be nonzero. Furthermore, $k_{0,0} = 1/2 - k_{0,1} = 1/2 - k_{1,0} = k_{1,1}$, so only one of these numbers needs to be determined. A careful study of the possible values of the $V_1, \ldots, V_R$ variables shows that

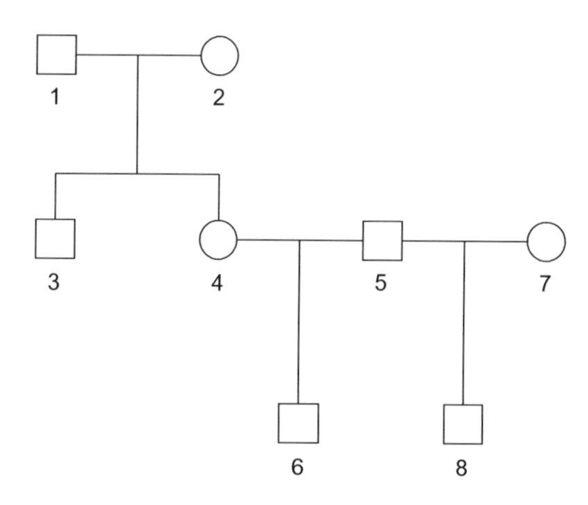

FIGURE 6.6

Pedigree for Section 6.4.1.

$$k_{1,1} = \frac{1}{2}(1-\rho) \qquad \text{for } \mathcal{P}_1, \tag{6.20}$$

$$k_{1,1} = \frac{1}{2}(\rho^2 + (1-\rho)^2) \qquad \text{for } \mathcal{P}_2, \tag{6.21}$$

$$k_{1,1} = \frac{1}{2}((1-\rho)(\rho^2 + (1-\rho)^2) + \rho/2) \qquad \text{for } \mathcal{P}_3. \tag{6.22}$$

See Thompson [93, pp. 59-60] for further details.

The LR comparing any pair of the three pedigrees can now be computed with Equation 6.17, and will generally be different from 1. The situation is particularly simple if the individuals are heterozygous for different alleles for both markers as then $\Pr(\overline{\mathcal{G}} \mid \overline{K}) = 0$, except when $\overline{K} = \sigma_{0,0}$, as no allele sharing is possible in this case. The LR will reduce to the quotient of the values of $k_{0,0}$ corresponding to the two pedigrees. Specifically, the grandparent versus half sibling comparison leads to

$$\text{LR} = \frac{1-\rho}{\rho^2 + (1-\rho)^2}. \tag{6.23}$$

Exercise 6.6 asks for a numerical check of this LR.

A simulation study designed to study the possibility of distinguishing the symmetric pedigrees mentioned is included in [103]. As always, the power depends on the markers, more SNPs than STRs are required. The basic conclusion is, however, that one will need a large number of pairs of linked markers to obtain reliable conclusions.

6.5 EXERCISES

Exercise 6.1 Mutation matrix. Show that the matrix specified in Equation 6.12 is a mutation model (i.e., it satisfies $M\mathbf{1}^t = \mathbf{1}^t$ and that all its entries are nonnegative), that it is stationary, and that its expected mutation rate is R. Note that the expected mutation rate may be computed as $1 - \text{Tr}(D(p)M)$, where $D(p)$ means the $n \times n$ diagonal matrix with p along its diagonal, and Tr is the trace operator.

Exercise 6.2 Mixtures and relatives. We consider some calculations relevant for the examples in Kaur et al. [75] based on `Familias`. This is an example of a relationship problem combining DNA mixture data and individual reference profiles. We consider only one marker. There is a mixture $E = 1/2/3$ from a mother and a child (fetus). Genotype data are only available from the alleged father, and he is 3/4. We consider hypothesis H_1 (the alleged father is the biological father of the child) versus hypothesis H_2 (a man unrelated to the mother and the alleged father is the father). In this exercise, we will allow there to be mutations.

(a) Use the `generate` function of `BookEKM`[8] to generate a data matrix—that is, all combinations of mother and child consistent with the mixture. As mutations

[8] Available from http://familias.name/book.html.

Table 6.1 The Data Matrix for Exercise 6.2

CH	2/3	1/3	1/2	3/3	1/3	2/3	2/2	1/2	2/3	1/1	1/2	1/3
MO	1/1	2/2	3/3	1/2	1/2	1/2	1/3	1/3	1/3	2/3	2/3	2/3
AF	3/4	3/4	3/4	3/4	3/4	3/4	3/4	3/4	3/4	3/4	3/4	3/4

Abbreviations: AF, alleged father; CH, child; MO, mother.

will be accommodated, also combinations without allele sharing between the mother and the child are included. Add the genotype AF to the data matrix. Your data should be a data frame with three rows and 24 columns of numbers as given in Table 6.1.

(b) The alleles are 1, 2, 3, 4, and 5 with uniform allele frequencies. Use the default mutation model with a mutation rate R for males and females of 0.005. The data matrix can be viewed as representing genotype data for 12 loci. Use `mix3Familias` to find the LR. It can be shown that $\text{LR} = 1/(6p_3)$ if the mutation rates are set to 0. Use this to verify the calculations above for the case without mutations.

(c) Let $R = 0$ and calculate LR(θ) for values of θ specified by `seq(0, 0.1, length = 100)`

(d) Repeat (c) with $R = 0.005$. Make a plot to illustrate the result.

(e) Assume the mother is genotyped and is 1/2. Calculate the LR by hand (assume there is no mutation and theta correction).

Exercise 6.3 Dropout rates. In Example 6.9, a simple model for dropouts was described, where each allele in the genotype has an independent probability d of not being observed in the data. As mentioned, when a heterozygote is observed, dropouts cannot have occurred, while if the data are from a homozygote, there is a certain probability that a dropout has occurred. However, it is important to notice that this probability is not equal to the probability d.

(a) Assume one has made the observation $a/-$ for a marker in which the allele a has population frequency p_a. Let D be a variable which has possible values 0, 1, and 2 according to whether there has been zero, one, or two dropouts at the marker. Use Bayes's formula to compute the formula for $\Pr(D = 1 \mid \text{data} = a/-)$ in terms of d and p_a. What will happen to this probability when p_a approaches zero?

(b) When the observation is $a/-$, there are two possibilities for the actual genotype: that it is homozygous a/a, or that it is heterozygous a/b, with $b \neq a$. Use Bayes's formula to compute a formula for the probability $\Pr(\text{heterozygote} \mid \text{data} = a/-)$ in terms of d and p_a.

Exercise 6.4 *Inbreeding. Jacquard. `paramlink`. In this exercise, we look at *Jacquard condensed identity states* $\Sigma_1, \ldots, \Sigma_9$, which describe the possible IBD patterns between two individuals; see Section 6.4.1. These states are illustrated in the second row in Figure 6.5. The names of the states are given in the first row in Figure 6.5.

Given a pedigree connecting two persons, it will induce a probability distribution on the nine possible identity states. We defined

$$\Delta_i = \Pr(\Sigma_i \mid \mathcal{P})$$

in Section 6.4.1 for a pedigree $\mathcal{P}$. The `R` functions below are found in the package `paramlink`.

(a) Explain why $\Delta_1 = \cdots = \Delta_6 = 0$ when individuals X and Y are not inbred.

We now assume our marker has two alleles, a and b, with population frequencies p and $q = 1 - p$, respectively. Any individual then has one of the three genotypes $g_1 = a/a$, $g_2 = a/b$, and $g_3 = b/b$. For a pair of individuals X and Y with genotypes G_X and G_Y there are then nine possible combinations of genotypes. Given a pedigree $\mathcal{P}$ and the alleles with population frequencies above, we get a probability distribution on these possible genotypes. Let $d(\mathcal{P})$ denote the 3×3 matrix with entries $d_{ij}(\mathcal{P}) = \Pr(G_X = g_i, G_Y = g_j \mid \mathcal{P})$. We then get

$$\begin{aligned} d_{ij}(\mathcal{P}) &= \sum_{k=1}^{9} \Pr(G_X = g_i, G_Y = g_j, \Sigma_k \mid \mathcal{P}) \\ &= \sum_{k=1}^{9} \Pr(G_X = g_i, G_Y = g_j \mid \Sigma_k) \Pr(\Sigma_k \mid \mathcal{P}) = \sum_{k=1}^{9} d_{ij}(\Sigma_k)\Delta_k. \end{aligned}$$

(b) For noninbred pairs of individuals, the following three distributions are particularly important:

$$\begin{aligned} \text{UN} &:= d(\Sigma_9) = d(\text{X and Y are unrelated}), \\ \text{PO} &:= d(\Sigma_8) = d(\text{X and Y are parent and offspring}), \\ \text{MZ} &:= d(\Sigma_7) = d(\text{X and Y are monozygotic twins}). \end{aligned}$$

Assume X and Y are full siblings. Verify that

$$d(\mathcal{P}) = \tfrac{1}{4}\text{UN} + \tfrac{1}{2}\text{PO} + \tfrac{1}{4}\text{MZ}$$

and that

$$d_{1,1}(\mathcal{P}) = \tfrac{1}{4}p^2(p^2 + 2p + 1). \tag{6.24}$$

The remainder of this exercise addresses a problem involving an inbred pairwise relationship. The hypotheses are as follows:

H_1: The individuals are full brothers whose parents are full siblings.
H_2: The individuals are full brothers with unrelated parents.

For H_1 (see below),

$$\Delta = \left(\frac{2}{32}, \frac{1}{32}, \frac{4}{32}, \frac{1}{32}, \frac{4}{32}, \frac{1}{32}, \frac{7}{32}, \frac{10}{32}, \frac{2}{32}\right), \tag{6.25}$$

while for H_2,

$$\Delta = \left(0, 0, 0, 0, 0, 0, \frac{1}{4}, \frac{1}{2}, \frac{1}{4}\right). \tag{6.26}$$

Assume both individuals are homozygous a/a.

(c) Create and plot the pedigrees corresponding to H_1 and H_2 with genotypes indicated.

(d) Let $p = 0.1$. Use `oneMarkerDistribution` to verify that the distribution for the genotypes if H_1 holds is as given in Table 6.2.

(e) Show that the likelihood for H_1 is

$$\Pr(\text{data} \mid H_1) = \frac{1}{16}(1 + 8p + 6p^2 + p^3)p.$$

(f) Derive

$$\text{LR}(p) = \frac{\Pr(\text{data} \mid H_1)}{\Pr(\text{data} \mid H_2)} = \frac{1}{4}\frac{1 + 8p + 6p^2 + p^3}{p + 2p^2 + p^3}. \tag{6.27}$$

Verify that $\lim_{p \to 0^+} \text{LR}(p) = \infty$, $\text{LR}(1) = 1$, and $\text{LR}(p) \geq 1$. Plot $\log_{10} \text{LR}(p)$ for $p \in [0.001, 0.1]$.

(g) Suppose one of the brothers is a/a; the other is not typed. Find the conditional probability for the genotypes of the other brother with `oneMarkerDistribution`. Also explain how the result could have been obtained from Table 6.2.

(h) The function `markerSim` implements simulation allowing for general conditioning and complex pedigrees. Use `markerSim` to check the exact calculations above.

Exercise 6.5 Jacquard coefficients. `identity`.

(a) Use the `R` package `identity` to calculate the condensed identity coefficients Δ given in Equations 6.25 and 6.26.

(b) A father has a son by his own daughter. Find Δ and calculate the likelihood when the father and son are both a/a for a SNP marker with frequency $p = 0.1$ for the a allele with use of Equation 6.24. Use `Familias` or `paramlink` to confirm the answer.

Table 6.2 Table for Exercise 6.4

	a/a	*b/b*	*a/b*
a/a	0.0116	0.0061	0.0147
b/b	0.0061	0.7756	0.0507
a/b	0.0147	0.0507	0.0695

Exercise 6.6 *IBD matrix. (LR for linked markers). The purpose of this exercise is to calculate and plot the LR comparing full siblings with unrelated individuals for two linked markers.

(a) Estimate the matrix defined by Equation 6.19 for markers 10 cM apart and store the result as a 3×3 matrix `kappa`. *Hint*: See the documentation for `twoLocusIBD` in the `R` package `BookEKM`.

(b) Assume next both brothers are 1/1 for both markers for a SNP marker with alleles 1 and 2 with frequencies p and $1 - p$. The commands below calculate the LR comparing brothers with unrelated individuals for $p \in [0.01, 1]$:

```
pp <- seq(0.01, 1, length = 100)
L.brothers <- NULL
for (p in pp){
  d1 <- d2 <- c(p^4,p^3,p^2)
  L.brothers <- c(L.brothers, t(d1)%*% kappa %*% d2)
}
LR <- L.brothers/pp^8
```

Plot LR (on $\log_{10}$ scale) as function of p.

CHAPTER

Parameter estimation and uncertainty

7

CHAPTER OUTLINE

The main focus of this book is to compute the probability of marker data given a hypothesized pedigree, population-level, pedigree-level, and observational-level models, and parameters for these models. So far, we have just assumed particular values for the parameters. However, parameters for real models need to be estimated from data, and in practice parameters can never be known with complete accuracy. In this chapter, we study ways to estimate our model parameters. We also study ways to take into account in the final results the remaining uncertainty in parameter values.

Below, we mainly discuss allele frequencies, the theta-correction parameter, the λ parameter modeling haplotype frequencies, and mutation rates and models. However, we hope to convey how similar approaches can be applied to other parameters.

7.1 ALLELE FREQUENCIES

Our first example of how to deal with parameter estimation and uncertainty concerns allele frequencies. Our approach was described in Sections 6.1.1 and 6.1.2 and illustrates the general idea: First, one establishes a prior for the parameter in question, the vector $\mathbf{p}$ of allele frequencies. The prior is based on biological knowledge of the context, so, for example, alleles that are considered biologically possible are included in the model even when they have not been observed in the relevant database. We used a Dirichlet prior for $\mathbf{p}$. The posterior for $\mathbf{p}$ given a database with counts of alleles was then found to also be a Dirichlet distribution. In Section 6.1.2 we saw that, if possible, likelihoods depending on parameters for which we have a (posterior) probability distribution should be computed by marginalizing over the parameter of the joint distribution of the data and the parameter. This will generally give different answers than computing with some fixed estimate for the parameter. Regarding allele frequencies, one should thus compute the likelihood ratio (LR) as follows:

$$\mathrm{LR} = \frac{\int_{\mathbf{p}} \Pr(\mathcal{D} \mid \mathbf{p}, \mathcal{P}_1)\pi(\mathbf{p})\, d\mathbf{p}}{\int_{\mathbf{p}} \Pr(\mathcal{D} \mid \mathbf{p}, \mathcal{P}_2)\pi(\mathbf{p})\, d\mathbf{p}}.$$

Instead of using the above approach to compute the LR, one may be tempted to first compute how the LR depends on the parameter in question, and then marginalize over the parameter of the joint distribution of the LR and the parameter; in other words, compute

$$\int_{\mathbf{p}} \frac{\Pr(\mathcal{D} \mid \mathbf{p}, \mathcal{P}_1)}{\Pr(\mathcal{D} \mid \mathbf{p}, \mathcal{P}_2)} \pi(\mathbf{p})\, d\mathbf{p}.$$

However, this will generally give a different, answer, as we will see in the next example.

Example 7.1 *Uncertainty in LR. Let us consider a standard duo paternity case, where both the alleged father and the child have genotype A/A. We let $\mathcal{P}_1$ and $\mathcal{P}_2$ denote the pedigrees of paternity and nonpaternity, respectively. We assume the allele frequency vector $\mathbf{p}$ has a Dirichlet distribution with parameter vector λ with sum Λ. Thus, each λ_i is the sum of the counts of allele i in the database and whatever pseudocount for this allele is used, while Λ is the number of alleles in the database plus the sum of all the pseudocounts. Using $(\ldots, 0, 0, 3, 0, 0, \ldots)$ and $(\ldots, 0, 0, 4, 0, 0, \ldots)$ to denote the vectors of counts that indicate three or four counts of allele A, respectively, using Equation 6.4, we get (ignoring mutations)

$$\begin{aligned}
\mathrm{LR} &= \frac{\Pr(\mathcal{D} \mid \mathcal{P}_1)}{\Pr(\mathcal{D} \mid \mathcal{P}_2)} = \frac{\int_{\mathbf{p}} \Pr(\mathcal{D} \mid \mathbf{p}, \mathcal{P}_1)\pi(\mathbf{p})\, d\mathbf{p}}{\int_{\mathbf{p}} \Pr(\mathcal{D} \mid \mathbf{p}, \mathcal{P}_2)\pi(\mathbf{p})\, d\mathbf{p}} \\
&= \frac{\int_{\mathbf{p}} p_A^3 \pi(\mathbf{p})\, d\mathbf{p}}{\int_{\mathbf{p}} p_A^4 \pi(\mathbf{p})\, d\mathbf{p}} = \frac{B(\lambda + (\ldots, 0, 0, 3, 0, 0, \ldots))/B(\lambda)}{B(\lambda + (\ldots, 0, 0, 4, 0, 0, \ldots))/B(\lambda)} \\
&= \frac{\Gamma(\lambda_A + 3)/\Gamma(\Lambda + 3)}{\Gamma(\lambda_A + 4)/\Gamma(\Lambda + 4)} = \frac{\Lambda + 3}{\lambda_A + 3}.
\end{aligned}$$

On the other hand, if we assume $\lambda_A > 1$, we get

$$\begin{aligned}
\int_{\mathbf{p}} \frac{\Pr(\mathcal{D} \mid \mathbf{p}, \mathcal{P}_1)}{\Pr(\mathcal{D} \mid \mathbf{p}, \mathcal{P}_2)} \pi(\mathbf{p})\, d\mathbf{p} &= \int_{\mathbf{p}} \frac{p_A^3}{p_A^4} \pi(\mathbf{p})\, d\mathbf{p} = \int_{\mathbf{p}} p_A^{-1} \pi(\mathbf{p})\, d\mathbf{p} \\
&= \frac{B(\lambda + (\ldots, 0, 0, -1, 0, 0, \ldots,))}{B(\lambda)} = \frac{\Gamma(\lambda_A - 1)/\Gamma(\Lambda - 1)}{\Gamma(\lambda_A)/\Gamma(\Lambda)} \\
&= \frac{\Lambda - 1}{\lambda_A - 1}.
\end{aligned}$$

The conventional answer, disregarding uncertainty, is $1/p_A = \Lambda/\lambda_A$. So when the count λ_A is fairly large, the difference between the results will be slight. However, there will be a difference. As we have discussed, integrating the numerator and denominator of the likelihood separately will produce an LR taking uncertainty into account, so the second way of thinking should be avoided.

7.2 *THE THETA-CORRECTION PARAMETER

In Sections 2.5 and 6.1.3, we saw how the parameter θ may be used in a way that models population substructure, or "kinship," or "inbreeding." To implement computations based on this we need to estimate a reasonable range of values for θ, for various situations. A number of possibilities exist, depending on what kind of data is available. In this book we consider only one possible method below, but see also [40, 104]. In [13, page 73], based on [105], estimation of the inbreeding parameter is explained for a diallelic locus leading to a posterior.

Assume we have a large database from which we obtain allele frequencies for a general population. In this context, we will assume the database is large enough so that we can ignore uncertainty in these allele frequencies, and we use a fixed estimate $\hat{\mathbf{q}}$. Assume also that we have a smaller sample from some subpopulation, with a vector $\mathbf{c}$ of counts of alleles, with a total count of n. In Section 6.1.3, we saw that the use of a kinship parameter θ to represent the degree of separation of this subpopulation from the general population corresponds to the use of a Dirichlet distribution for the subpopulation allele frequencies $\mathbf{p}$ with parameter

$$\lambda = (1/\theta - 1)\hat{\mathbf{q}}.$$

This means that we can now compute a likelihood for the data $\mathbf{c}$ given various values of the parameter θ: The probability of $\mathbf{c}$ given $\mathbf{p}$ is multinomial,

$$\Pr(\mathbf{c} \mid \mathbf{p}) = \frac{n!}{c_1!c_2!\dots c_k!} p_1^{c_1} p_2^{c_2} \dots p_k^{c_k}.$$

Using Equation 6.4, we get

$$\begin{aligned}
\Pr(\mathbf{c} \mid \theta) &= \int_{\mathbf{p}} \Pr(\mathbf{c} \mid \mathbf{p}) \pi(\mathbf{p} \mid \theta)\, d\mathbf{p} \\
&= \frac{n!}{c_1!c_2!\dots c_k!} \int_{\mathbf{p}} p_1^{c_1} p_2^{c_2} \dots p_k^{c_k} \pi(\mathbf{p} \mid \theta)\, d\mathbf{p} \\
&= \frac{n!}{c_1!c_2!\dots c_k!} \times \frac{B(\lambda + \mathbf{c})}{B(\lambda)} \\
&= \frac{n!}{c_1!c_2!\dots c_k!} \times \frac{B((1/\theta - 1)\hat{\mathbf{q}} + \mathbf{c})}{B((1/\theta - 1)\hat{\mathbf{q}})} \\
&\propto_\theta \frac{B((1/\theta - 1)\hat{\mathbf{q}} + \mathbf{c})}{B((1/\theta - 1)\hat{\mathbf{q}})}.
\end{aligned} \tag{7.1}$$

In the last step, we discard the factors that do not depend on θ, as we are interested only in this as a likelihood function for θ. We could simplify the last expression further, as in Section 6.1.2, but for computational implementation in R, the above form is quite okay.

The function for $\Pr(\mathbf{c} \mid \theta)$ indicates which values of θ make the observed data $\mathbf{c}$ most likely. In real examples, we may combine this with a prior on θ to produce a posterior distribution. Below is a small toy example of how this can work.

Example 7.2 Estimation of kinship. Assume we have derived a population frequency estimate vector $\hat{\mathbf{q}} = (0.11, 0.21, 0.33, 0.35)$ from a large database. In a subpopulation we have observed the counts $\mathbf{c} = (8, 2, 2, 4)$. Thus, the counts indicate that the allele frequencies of the subpopulation are substantially different from those of the main population.

In the code below, we compute the probability of the counts $\mathbf{c}$ for 999 values of θ in a grid. If we use a flat prior for θ, these probabilities are proportional to the posterior distribution for θ, which we can then find by rescaling the vector of values. Figure 7.1 shows that there is a great deal of uncertainty about the value θ. The value of θ with the highest likelihood is 0.151. *Note*: θ values in realistic examples will often be much smaller than this.

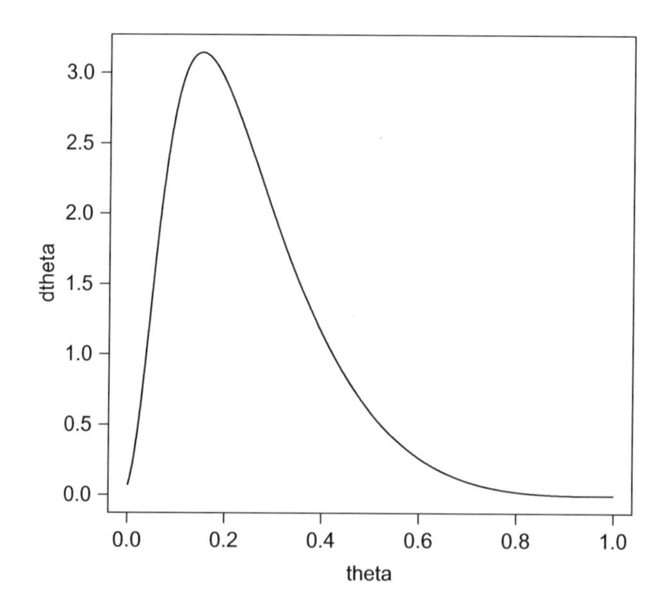

FIGURE 7.1

Output from Example 7.2.

```
# Define the data
p0 <- c(0.11, 0.21, 0.33, 0.35)
c <- c(8, 2, 2, 4)

# Create the grid for the likelihood function
theta <- (1:999)/1000
dtheta <- rep(0, 999)

#Define the logged multivariate Beta function
lMultivariateBeta <- function(alpha)
            sum(lgamma(alpha)) - lgamma(sum(alpha))

# Compute the likelihood values
for (i in 1:999)  dtheta[i] <- exp(
          lMultivariateBeta((1/theta[i]-1)*p0+c) -
          lMultivariateBeta((1/theta[i]-1)*p0))

# Create plot
dtheta <- dtheta/sum(dtheta)*1000
plot(theta, dtheta, type="l")

# Compute the most likely theta value
print(theta[which(dtheta==max(dtheta))])
```

7.2.1 TAKING THETA UNCERTAINTY INTO ACCOUNT

The θ parameter is difficult to estimate, whether one uses the procedure in the previous subsection or some other approach. In most cases, substantial uncertainty will remain, which raises the issue of how this uncertainty can be taken into account in likelihood computations.

As described at the beginning of Section 6.1.2, the best way to deal with the uncertainty is to integrate it out. However, unlike for allele frequencies, this cannot be done analytically for the kinship parameter, and a numerical approach must be used instead. Below, we show an alternative which uses simplistic numerical integration. If a vector containing values for a density function for θ is available, one may use such a vector together with repeated computations of the likelihoods to obtain a good approximation of the correct result.

Example 7.3 Taking uncertainty in θ into account. Our example builds on Example 7.2: We assume that we have a vector `theta` of investigated values for θ and a vector `dtheta` of probability densities at these values, as were produced in that example. We then assume we have some relationship inference problem where the case data is assumed to come from a subpopulation with similar amounts of kinship as was estimated in Example 7.2. In the code below, we compute the LR without kinship, the LR with kinship at the likeliest value of θ, and the LR when we integrate out the uncertainty in θ. The results are 1.67, 1.79, and 1.78, respectively. The last one is arguably the one to trust. *Note*: The amount of kinship, as estimated in Example 7.2, is larger than in most realistic examples.

```
library(Familias)
persons <- c("mother", "child", "AF")
sex <- c("female", "male", "male")
ped1 <- FamiliasPedigree(id = persons, dadid = c(NA, NA, NA),
momid = c(NA, "mother", NA), sex=sex)
ped2 <- FamiliasPedigree(id = persons, dadid = c(NA, "AF", NA),
momid = c(NA, "mother", NA), sex=sex)
pedigrees <- list(notFather= ped1, isFather= ped2)
frequencies <- c(0.1, 0.2, 0.3, 0.4)
allelenames <- c("A", "B", "C", "D")
marker <- FamiliasLocus(frequencies, allelenames)
mother <- c("A", "B")
child <- c("A", "C")
AF <- c("C", "D")
datamatrix <- rbind(mother, child, AF)

# Compute result with theta = 0:
result <- FamiliasPosterior(pedigrees, marker, datamatrix)
print(result$LR["isFather"])

# Compute result with theta = 0.151:
result2 <- FamiliasPosterior(pedigrees, marker, datamatrix,
  kinship = 0.151)
print(result2$LR["isFather"])

# Compute result integrating out over values of theta:
n <- length(theta)
tmp <- matrix(0, n, 2)
for (i in 1:n) tmp[i,] <- FamiliasPosterior(pedigrees, marker,
   datamatrix, kinship = theta[i])$likelihoods * dtheta[i]
tmp <- 0.5*(tmp[-n,]+tmp[-1,])
result3 <- apply(tmp * (theta[-1]-theta[-n]), 2, sum)
print(result3[2]/result3[1])
```

7.3 THE LAMBDA MODEL FOR HAPLOTYPE FREQUENCIES

In Sections 4.3 and 6.1.4 we discussed the λ model for estimating haplotype frequencies. As mentioned, the resulting haplotype frequency estimates may depend quite a bit on the value of λ. Thus, we now consider how such a parameter may be estimated.

Consider a specific situation, where we have a relatively small database of haplotypes and want to estimate haplotype probabilities from this database. Assume allele frequencies have been computed from another database. If we want to use the λ method above together with these data, what value for λ should we use? The value of λ will in some sense indicate how much linkage disequilibrium (LD) we expect there to be in the haplotype database. When choosing a large λ, we expect little LD: a large λ means that the prior distribution for the haplotype probabilities has a fairly small uncertainty around the estimate where each haplotype frequency equals the product of the frequencies of its alleles. On the other hand, choosing a small λ means that we are more open to the possibility that there might be large amounts of LD. A way to better understand the relationship between λ and concrete examples of LD is to use specific haplotype data to estimate λ, in a way similar to that in which θ was estimated above.

Example 7.4 *Estimation of λ. Let us look at two markers, where we have estimated allele frequencies `freq1` and `freq2` from a large database. Let $\mathbf{F}$ be the vector of expected frequencies of haplotypes under linkage equilibrium, derived as the products of all possible combinations of the frequencies in `freq1` and `freq2`. Then we use as a prior for the haplotype frequencies a Dirichlet distribution with parameter vector $\lambda\mathbf{F}$. Assume we have a small database with n haplotypes, with counts of the different haplotypes given in the vector $\mathbf{c}$. In a way similar to Equation 7.1, the probability of this data for each value of λ can be computed as

$$\Pr(\mathbf{c} \mid \lambda) = \frac{n!}{c_1!c_2!\dots c_k!} \times \frac{B(\lambda\mathbf{F}+\mathbf{c})}{B(\lambda\mathbf{F})}.$$

Below, we study the log-likelihood as a function of $\log(\lambda)$; we may then ignore the first factor on the right-hand side above as it does not depend on λ. We look at two different constructed examples of haplotype databases: In the first example, we simulate haplotypes as if there is no LD. *Note*: Rerunning the code may produce tables different from the ones below, as this is a simulation. In the second example, we adjust these data to indicate some LD.

```
freq1 <- c(0.1, 0.3, 0.6)
freq2 <- c(0.1, 0.2, 0.3, 0.4)
F <- freq1%o%freq2
F
```

```
     [,1] [,2] [,3] [,4]
[1,] 0.01 0.02 0.03 0.04
[2,] 0.03 0.06 0.09 0.12
[3,] 0.06 0.12 0.18 0.24
```

```
data1 <- matrix(rmultinom(1, 100, F), 3, 4)
data1
```

```
     [,1] [,2] [,3] [,4]
[1,]    1    2    3    5
```

```
[2,]    0    5    9   11
[3,]    6   12   20   26
```

```
data2 <- data1; data2[2:3,3:4] <- data2[2:3,3:4]+
c(10, -10, -10, 10)
data2
```

```
     [,1] [,2] [,3] [,4]
[1,]    1    2    3    5
[2,]    0    5   19    1
[3,]    6   12   10   36
```

```
lMultivariateBeta <- function(alpha)
               sum(lgamma(alpha)) - lgamma(sum(alpha))
loglik <- function(data, lambda, base)
              lMultivariateBeta (lambda*base + data) -
              lMultivariateBeta (lambda*base)
lambda <- exp(seq(log(1),log(10000), length.out=1000))
likelihood <- lambda
for (i in 1:1000) likelihood[i] <-
              exp(loglik (data1, lambda[i], F))
plot(lambda, likelihood, log="x", type="l")
for (i in 1:1000) likelihood[i] <-
              exp(loglik (data2, lambda[i], F))
plot(lambda, likelihood, log="x", type="l")
```

Figure 7.2 shows how the likelihood of λ increases as λ increases, as the data is indeed from a model with haplotype frequencies given as products over allele frequencies. Figure 7.3 shows how the likelihood of λ reaches a maximum around $\lambda = 30$: for higher values, the resulting model does not really fit the data very well, as the data contains clear amounts of LD.

7.4 MUTATIONS AND MUTATION MODELS

As our third example of parameter estimation and uncertainty we consider mutation models and mutation model parameters. Similarly to our previous examples, mutation parameters may have a large influence on the LR. The issue is not so much estimating the probability that a mutation will occur, but rather estimating the probability of a specific mutation event, from a given allele to another given allele. We saw in Section 6.2.1 how different mutation models may have the same mutation rates but very different probabilities for specific mutation events.

Thus, learning about the entire mutation matrix, or deciding which mutation model to use, is as important as estimating mutation rates. Below, we will indicate in a simple example how one can learn about mutation rates from data. Choosing between mutation models based on data may in fact be done in a similar way.

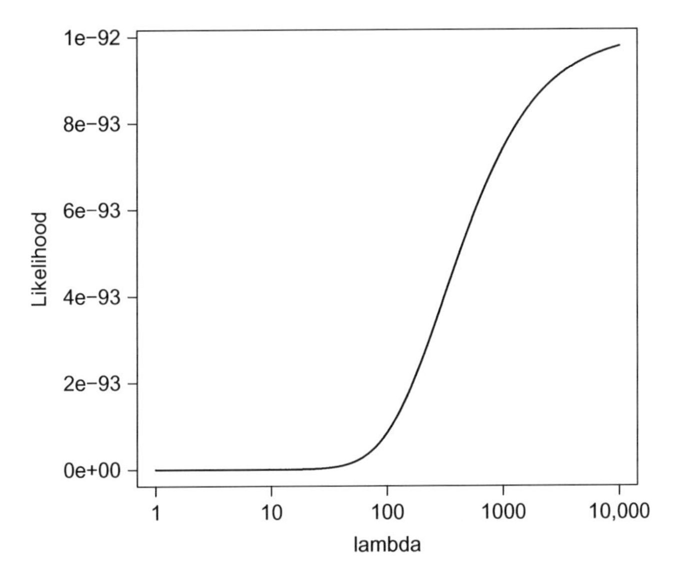

FIGURE 7.2

The likelihood of λ for data simulated without LD.

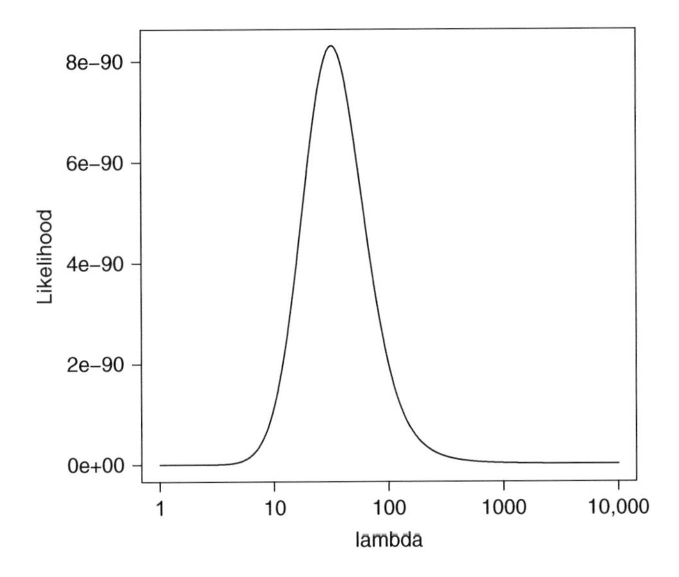

FIGURE 7.3

The likelihood of λ for data where some LD has been added.

One may be tempted to estimate mutation rates for a locus by considering data from a number of cases where the true pedigree has been established with very little remaining uncertainty, and for a specific relationship count the proportion of times there is an exclusion in that relationship—that is, where the data can be explained only by a mutation. However, not all mutations lead to inconsistencies. Unless this is accounted for, the resulting estimates may be biased, as exemplified in [102] based on data from http://www.aabb.org/sa/facilities/Documents/rtannrpt08.pdf. The paper mentioned discusses models for estimation of mutation parameters.

Example 7.5 *Mutation rate estimation. We look at a toy example with simulated data. We imagine a situation with data from a single locus and 10,000 duo cases, where we have concluded from other data, with negligible risk of error, that there is true paternity. We show how to use such data to estimate the mutation rate, under the assumption of a specific fixed mutation model. We simulate three data sets, with mutation rates of 0, 0.003, and 0.006, respectively. There are four alleles and the frequencies are 0.1, 0.2, 0.3, and 0.4. Figure 7.4 shows likelihood curves for the data sets. We see that mutation rates of these sizes can be reasonably well

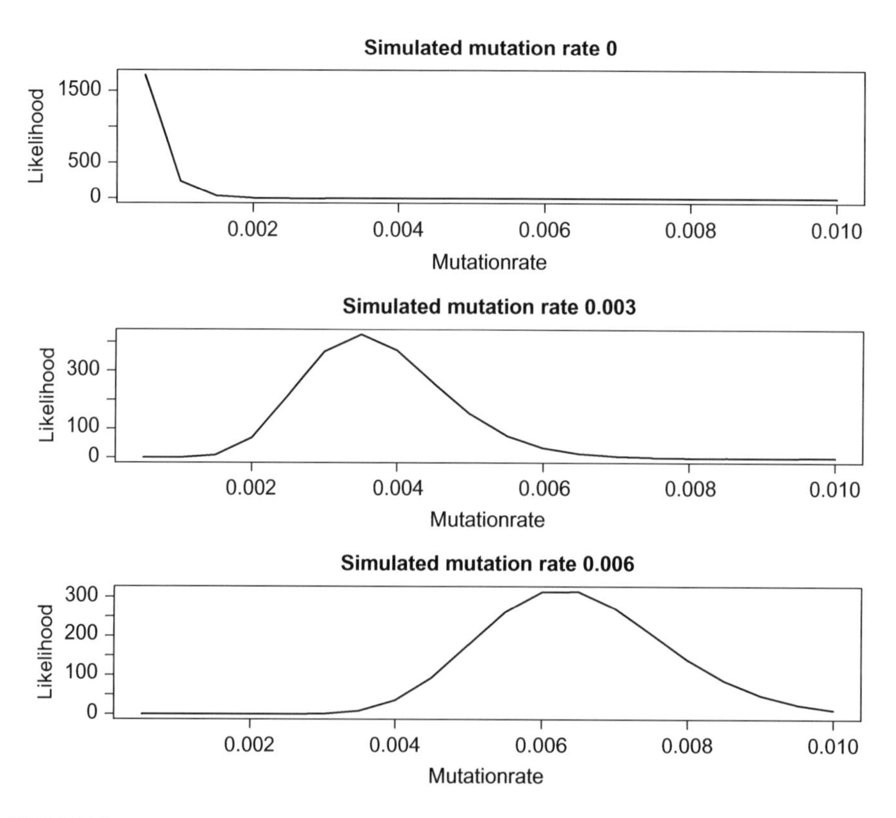

FIGURE 7.4

Likelihood curves for mutation rates based on three data sets, simulated with mutation rates of 0, 0.003, and 0.006, respectively.

estimated from 10,000 cases; note however that the main reason our results look nice is that we have simulated data with the same mutation model as is used in the likelihood computations (a "Stepwise" model with default range parameter 0.5). It is also possible to combine likelihood curves with prior distributions based on other, similar markers.

The data set simulated with a mutation rate of 0.003 ended up with having mutations in 34 of 10,000 cases. However, only 14 of these led to an exclusion. The corresponding numbers for the data set simulated with a mutation rate of 0.006 were 52 and 25. This illustrates how the use of counts of exclusions can easily lead to underestimated mutation rates (running the code below may take several minutes).

```
library(Familias)
simulateDuoData <- function(N, frequencies, allelenames,
  mutationrate) {
   ind    <- array(sample(length(allelenames), 4*N, replace=T,
                 p=frequencies), dim=c(2,2,N))
   result <- array(allelenames[ind], dim=c(2,2,N))
   mutmatrix <- FamiliasLocus(frequencies, allelenames =
      allelenames,
      MutationRate = mutationrate)$maleMutationMatrix
   for (i in 1:N) result[2,1,i] <- sample(allelenames, 1,
      prob = mutmatrix[ind[1,1,i],])
      print(sum(result[2,1,]!=result[1,1,])/N)
   result
}

loglikDuo <- function(mutationrate, data, frequencies,
   allelenames) {
   ped <- FamiliasPedigree(id = c("AF", "child"), dadid =
     c(NA, "AF"), momid = c(NA, NA), sex=c("male", "female"))
   n <- length(mutationrate); N <- dim(data)[3]
   res <- matrix(0, n, N)
   result <- rep(0, n)
   for (i in 1:n) {
      locus <- FamiliasLocus(frequencies, allelenames =
        allelenames, MutationRate = mutationrate[i])
      for (j in 1:N) {
         datamatrix <- data[,,j]
         rownames(datamatrix) <- c("AF", "child")
         res[i,j] <- FamiliasPosterior(ped, locus,
           datamatrix)$likelihoods[1]
      }
      result[i] <- sum(log(res[i,]))
   }
   result - mean(result)
}
```

```
plotlikelihood <- function(simulateMutationRate, frequencies,
   allelenames) {
   mutationrate <- seq(0.0005, 0.01, 0.0005)
   data <- simulateDuoData(10000, frequencies,
      allelenames, simulateMutationRate)
   exclusion <- (data[1,1,]!=data[2,1,] & data[1,1,]
                !=data[2,2,] &data[1,2,]
                !=data[2,1,] & data[1,2,]
                !=data[2,2,])
   print(sum(exclusion)/10000)
   result <- exp(loglikDuo(mutationrate, data, frequencies,
     allelenames))
   plot(mutationrate, result/sum(result)/0.0005,
      type = "l", ylab="likelihood", main=paste("Simulated
        mutation rate", simulateMutationRate))
}

frequencies <- c(0.1, 0.2, 0.3, 0.4)
allelenames <- 10:13
par(mfrow=c(3,1))
plotlikelihood(0, frequencies, allelenames)
plotlikelihood(0.003, frequencies, allelenames)
plotlikelihood(0.006, frequencies, allelenames)
```

7.5 OTHER PARAMETERS

Likelihood-based estimation can be applied to all the parameters of our models in ways similar to those in the examples above. The differences lie in what type of data carries information about the parameter. Recombination rates need to be inferred from data that depend on such rates. They are often well estimated from work concerned with genetic disease mapping.

Allele frequency rates for silent alleles may, in principle, be estimated from the apparent overabundance of homozygotes they create. But whether or not an allele is silent depends on the detection method, so it is much more powerful to compare detection methods to find silent alleles. Statistics on silent allele frequencies are updated at http://www.cstl.nist.gov/strbase/NullAlleles.htm.

Dropouts and silent alleles are somewhat similar, except of course that silent alleles are inherited in a pedigree, while dropouts are not. The dropout probability will also depend much on the circumstances of the measurement, so an estimate needs to be made for each particular circumstance. There are a number of relevant models and papers, including [106], aiming at mixture applications.

7.6 HANDLING "UNCERTAINTY" IN LRs

In our context, as well as in many other contexts where LRs are computed as part of a procedure to make decisions, people may want to discuss the uncertainty of an LR computed from specific case data. Such language is unfortunate, as such an LR is in itself not a random variable, and thus does not have an uncertainty associated with it. As we have seen, when there is uncertainty connected to a parameter θ involved in our models, the LR should be computed as

$$\mathrm{LR} = \frac{\int_\theta \Pr(E \mid \theta)\pi(\theta)\,\mathrm{d}\theta}{\int_\theta \Pr(E \mid \theta)\pi(\theta)\,\mathrm{d}\theta}.$$

This is different from estimating LR as $\Pr(E \mid \hat{\theta})/\Pr(E \mid \hat{\theta})$ where some fixed estimate $\hat{\theta}$ of θ is used, or integrating over the uncertainty in the parameter after the LR has been computed.

However, an important reason why it may be tempting to talk about "uncertainty in the LR" is that in some situations it is known that the LR depends heavily on parameters which are quite uncertain. For example, although the LR for a paternity case in which there are no apparent exclusions will be little influenced by parameters other than the allele frequencies of observed alleles, the LR for a paternity case with an apparent exclusion will depend heavily on the probabilities of particular mutations, and such probabilities are estimated only with considerable uncertainty.

There are, however, several problems with the use of a language where such LRs have uncertainty. One is that the LR is somehow viewed as a model parameter, and that one attempts to learn about the LR by combining information about it from several sources, in a Bayesian fashion. Another problem is that one may obscure the connection with the underlying parameter that is actually causing the apparent uncertainty in the LR.

Instead of using the language of "uncertainty in LRs," we recommend talking about how the LR depends on specific parameters. For example, one may study how reducing the uncertainty in a parameter changes the value of the LR.

One may also study the LR as a function of the data, before data is observed. In such a context, a stochastic model for the data turns the LR into a random variable, and we may study the properties of its distribution. This perspective will be discussed in Chapter 8.

7.7 EXERCISE

Exercise 7.1 Kinship by simulation. The code below defines a function which can be used to approximate kinship computations with use of simulation.

```
library(Familias)

# Define a function that makes a random sample
# from a Dirichlet distribution with parameter vector alpha.
```

```
rdirichlet <- function (alpha)
{
  v <- rgamma(length(alpha), alpha)
  v/sum(v)
}

# Define a function that computes kinship using simulation
kinshipBySimulation <-
function (pedigrees, loci, datamatrix, kinship, N = 10)
{
   if (class(loci)=="FamiliasLocus") {
      loci <- list(loci)
      nloci <- 1
   } else nloci <- length(loci)
   sresult <- matrix(0, nloci, length(pedigrees))
   sqrresult <- sresult
   newloci <- loci
   for (i in 1:N) {
     if (kinship > 0) for (j in 1:nloci)
       newloci[[j]]$alleles[] <-
       rdirichlet(loci[[j]]$alleles*(1/kinship - 1))
     res <-  FamiliasPosterior(pedigrees, newloci,
                  datamatrix)$likelihoods
     sresult <- sresult + res
     sqrresult <- sqrresult + res^2
   }
   myres <- sresult/N
   myerr <- sqrt((sqrresult - sresult^2/N)/(N*(N-1)))
   list(likelihoods = myres, standardErrors = myerr,
        approxmin = myres - 2*myerr, approxmax = myres
          + 2*myerr)
}
```

Consider a case where one investigates whether two persons are first cousins or unrelated. Imagine for simplicity one has data for the single marker D21S2055, and that both persons have genotype 28/28. Use frequencies from the `NorwegianFrequencies` of the `Familias` package, and assume a mutation rate of 0.005. Compute the LR with `Familias`. Compare the results obtained when kinship = 0 and kinship = 0.1. Then reproduce the result with kinship with use of the code above. Assess the accuracy and usefulness of the simulation method in this context.

CHAPTER

Making decisions

8

CHAPTER OUTLINE

In applied science, our goal can often be described as making various kinds of decisions optimally. So far in this book, we have mainly focused on the problem of deciding on which pedigree connects a set of genetically tested persons. But there are also many other questions arising in the context of DNA testing. Some examples are:

- Given some hypotheses about pedigrees connecting some persons, and plans to test them at certain markers, can we be reasonably sure the planned data will solve the case? Will it be necessary to test additional persons in the pedigree, or do tests at additional markers?
- In terms of our general ability to reach a conclusion, is this DNA-typing kit or technology better than another one?
- When there is a large uncertainty in some model parameters (e.g., population allele frequencies for some small population), how does the cost of collecting data to reduce uncertainty in the parameters compare with the resulting improvements in decisions?

Such questions can all be put into a decision-theoretic framework, so that computations can help answer them. In this chapter, we outline such a framework. We will then also be led to the study of the likelihood ratio (LR) as a stochastic variable—that is, the study of what we can predict about the LR before we have observed the data. Decision theory is described in a general framework in [107], while forensic applications are dealt with in [108].

8.1 SOME BASIC DECISION THEORY

Consider a situation where only two hypotheses, H_1 and H_2, are thought to be possible, and where probabilities $\Pr(H_1)$ and $\Pr(H_2) = 1 - \Pr(H_1)$ are available (usually computed with the help of some data). When a choice between the two must be made, one can make two errors: choosing H_1 when H_2 is true, and vice versa. Each of these errors can be imagined to result in some kind of "cost." We will see below that an optimal decision, one that results in minimal expected costs, will depend on comparing the ratio $\Pr(H_1)/\Pr(H_2)$ with the ratio of these costs. To form this ratio, we must assume that $\Pr(H_2){>}0$; however, when $\Pr(H_2) = 0$, there can be little doubt that we should choose H_1, so in the following we will always assume that $\Pr(H_2) > 0$.

Let us generalize to a situation where three choices are possible: one can decide for H_1, one can decide for H_2, or one can decide to stay undecided. We assume there is no cost if we make the correct decision, while the costs in the other cases are specified below:

$$\begin{array}{ll} H_1 \text{ is true but we decide for } H_2: & 1 + c_1, \\ H_2 \text{ is true but we decide for } H_1: & 1 + c_2, \\ \text{Make no decision:} & 1. \end{array}$$

Note that we will only need to concern ourselves with relative costs, so there is no limitation in assuming that the cost of no decision is 1. The expected cost when choosing H_1 is now $\Pr(H_2)(1 + c_2)$, while the expected cost when choosing H_2 is $\Pr(H_1)(1 + c_1)$. If we assume that $c_1c_2 > 1$, the principle of minimizing the expected cost tells us that we should

$$\begin{array}{ll} \text{choose } H_1 & \text{if } \Pr(H_2)(1 + c_2) < 1, \\ \text{choose } H_2 & \text{if } \Pr(H_1)(1 + c_1) < 1, \\ \text{make no decision} & \text{otherwise.} \end{array}$$

Observe that $c_1c_2 > 1$ implies that if $\Pr(H_2)(1 + c_2) < 1$, then $\Pr(H_1)(1 + c_1) > 1$ and vice versa; if $c_1c_2 \leq 1$, one will never choose to "make no decision."

Generally, our probabilities for the two hypotheses are based on information from data, (omitted in the notation below) often with use of a computation producing (posterior) odds:

$$o = \frac{\Pr(H_1)}{\Pr(H_2)} = \frac{\Pr(H_1)}{1 - \Pr(H_1)},$$

so $\Pr(H_1) = o/(o + 1)$ and $\Pr(H_2) = 1/(o + 1)$. Reformulating the decision rule in terms of the odds o, we get that we should

choose H_1	if $o > c_2$,
choose H_2	if $o < \frac{1}{c_1}$,
make no decision	otherwise.

DNA laboratories (or their customers) use test data to make the choice between hypotheses H_1 and H_2 (e.g., paternity or not paternity); let us call such data D. If $\Pr(D \mid H_2) = 0$, one will choose H_1, otherwise one computes $LR(D) = \Pr(D \mid H_1)/\Pr(D \mid H_2)$ and, using Bayes's formula in odds form and prior odds o_0, one obtains the posterior odds (presented previously in Section 2.3.5):

$$o = \mathrm{LR}(D) \times o_0.$$

One may then decide for either H_1 or H_2 by comparing o with cutoff values L_H and L_L. Specifically, one will

choose H_1	if $o > L_\mathrm{H}$,
choose H_2	if $o < L_\mathrm{L}$,
make no choice	otherwise.

Typically, L_H may be around 10,000 and L_L around 1/10,000. Some DNA laboratories may use $\mathrm{LR}(D)$ rather than the odds $o = \mathrm{LR}(D) \times o_0$ in the comparisons above. This may lead to suboptimal decisions; see Exercise 8.1. Although on the basis of their definitions at the start of this section it may seem difficult to assign numerical values to the relative costs c_1 and c_2, this is exactly what DNA laboratories do when they decide which cutoff values L_H and L_L to use, as clearly $L_\mathrm{L} = 1/c_1$ and $L_\mathrm{H} = c_2$.

8.1.1 EXCLUSIONS

Below we expand on Section 2.8 to put the *probability of exclusion* in a decision-theoretic setting. When we are comparing the probability of various data under two hypotheses H_1 and H_2, there may generally be data that has zero probability under H_1 but positive probability under H_2, and vice versa. However, $\mathrm{LR}(D) = \Pr(D \mid H_1)/\Pr(D \mid H_2)$ can be computed only for those D with $\Pr(D \mid H_2) > 0$. To keep $\mathrm{LR}(D)$ as our central concept, in this chapter we will restrict ourselves to models where any data with positive probability given H_1 also has positive probability given H_2. In our applications this will not be a serious restriction.

However, we do allow data D to have a positive probability given H_2 but zero probability given H_1. Let $\mathbb{E}$ be the set of all D such that $\Pr(D \mid H_1) = 0$. We can then define the *probability of exclusion* (PE) as

$$\mathrm{PE} = \Pr(\mathbb{E} \mid H_2).$$

So PE is the probability under H_2 of observing data that is impossible to observe under H_1.

In our types of applications, H_1 may be the hypothesis that a certain man "matches" a situation (in terms of being the real father or relative, or the person who deposited a crime scene trace), while H_2 may be the hypothesis that a "random man" from a relevant population matches. It is then relevant to define the value of *random man not excluded* (RMNE) as

$$\text{RMNE} = 1 - \text{PE}.$$

In other words, RMNE is the probability under H_2 that data will have a positive probability (i.e., not be excluded) under H_1.

When deciding between H_1 and H_2 on the basis of data D, one possibility is to decide for H_2 when $\Pr(D \mid H_1) = 0$ and $\Pr(D \mid H_2) > 0$, and otherwise stay undecided (or, if such situations could occur, decide for H_1 when $\Pr(D \mid H_2) = 0$ and $\Pr(D \mid H_1) > 0$). One then gets a decision rule that does not need specification of costs or cutoff values, and as such seems more "objective." However, its power depends on models where things are assumed to be "impossible,"—that is, have probability zero. In applications involving, for example, biology it tends to be dangerous, and at best a simplification, to consider things to be impossible. For example, apparent genetic mismatches may be the results of mutations, and mismatches between a trace and a genotype may be the result of measurement errors.

Example 8.1 Binary choice: Match, exclusion. Assume the data D is *binary*—that is, it can have only two values, 1 or 0; let us call these values "match" and "exclusion," respectively. Assume also that $\Pr(D = 0 \mid H_1) = 0$; in other words, when H_1 is true, one will never observe an exclusion, whereas one may observe a match or an exclusion when H_2 is true. We get

$$\text{LR}(D = 0) = \frac{\Pr(D = 0 \mid H_1)}{\Pr(D = 0 \mid H_2)} = 0,$$

$$\text{LR}(D = 1) = \frac{\Pr(D = 1 \mid H_1)}{\Pr(D = 1 \mid H_2)} = \frac{1}{\Pr(D = 1 \mid H_2)} > 1.$$

Thus, if $D = 0$, we get $o_0 \times \text{LR}(D) = 0 < 1/c_1$, and we should choose H_2. If $D = 1$, the assumption[1] $1/c_1 < o_0$ leads to

$$\frac{1}{c_1} < o_0 < o_0 \times \text{LR}(D = 1) = o,$$

so we should never choose H_2, while we should choose H_1 if

$$o_0 \times \text{LR}(D = 1) = \frac{o_0}{\Pr(D = 1 \mid H_2)} = \frac{o_0}{\text{RMNE}} > c_2.$$

[1] If $o_0 < 1/c_1$ we would have chosen H_2 already, on the basis of the prior information.

We can now sum up the situation in two alternatives: If the test is such that

$$\text{RMNE} < \frac{\text{prior odds}}{c_2},$$

one should choose H_1 if one observes a match and H_2 if one observes an exclusion. If RMNE is larger than (prior odds)$/c_2$, one should choose H_2 if one observes an exclusion, but should stay undecided if one observes a match.

When one uses a model with a nonzero PE, one may choose to use the data directly, or convert that data to "match" or "exclusion," and base the decision on that according to the example above. However, doing so may lead to suboptimal decisions, as we see in Exercise 8.2.

8.1.2 DECISIONS ABOUT ACQUIRING DATA

The framework above becomes more interesting when used to decide whether or not data should be acquired. Imagine as above that we have two hypotheses H_1 and H_2 and prior odds o_0 in favor of H_1, but that our decision is whether or not data D should be produced. We may assume that $L_L < o_0 < L_H$; otherwise we would have come to a conclusion based on the information already available. Given D we will be able to compute LR(D). The posterior odds given D will be LR(D) $\times$ o_0, and it is this number that should be compared with L_L and L_H to make a decision. The question is whether the cost of acquiring the data D is justified by the opportunities for reaching a correct conclusion given such data.

Defining c_D as the cost of acquiring D, we can express the expected cost after D has been acquired and analyzed as

$$\begin{aligned}
& c_D + \Pr(H_1)\left[\Pr\left(L_L \le \text{LR}(D) \times o_0 < L_H \mid H_1\right)\right. \\
& \quad \left. + \Pr\left(\text{LR}(D) \times o_0 < L_L \mid H_1\right)(1 + c_1)\right] \\
& \quad + \Pr(H_2)\left[\Pr\left(L_H > \text{LR}(D) \times o_0 \ge L_L \mid H_2\right)\right. \\
& \quad \left. + \Pr\left(\text{LR}(D) \times o_0 > L_H \mid H_2\right)(1 + c_2)\right] \\
& = c_D + \Pr(H_1)\left[\Pr\left(\text{LR}(D) \times o_0 < L_H \mid H_1\right) + \Pr\left(\text{LR}(D) \times o_0 < L_L \mid H_1\right) c_1\right] \\
& \quad + \Pr(H_2)\left[\Pr\left(\text{LR}(D) \times o_0 > L_L \mid H_2\right) + \Pr\left(\text{LR}(D) \times o_0 > L_H \mid H_2\right) c_2\right] \\
& = c_D + \frac{o_0}{o_0 + 1}\left[\Pr\left(\text{LR}(D) < \frac{L_H}{o_0} \mid H_1\right) + \Pr\left(\text{LR}(D) < \frac{L_L}{o_0} \mid H_1\right)\frac{1}{L_L}\right] \\
& \quad + \frac{1}{o_0 + 1}\left[\Pr\left(\text{LR}(D) > \frac{L_L}{o_0} \mid H_2\right) + \Pr\left(\text{LR}(D) > \frac{L_H}{o_0} \mid H_2\right) L_H\right].
\end{aligned}$$

The equation above may look ugly, but in fact its value can estimated. Let us define $\mathbb{X}_1$ and $\mathbb{X}_2$ as the random variables given by LR(D) when D has the distribution specified by H_1 or H_2, respectively.

We present further properties $\mathbb{X}_1$ and $\mathbb{X}_2$ later in this chapter. For most models discussed in this book, both $\Pr(D \mid H_1)$ and $\Pr(D \mid H_2)$ are positive for all D. Defining

$$V(o_0) = \frac{o_0}{o_0+1}\left[\Pr\left(\mathbb{X}_1 < \frac{L_H}{o_0}\right) + \Pr\left(\mathbb{X}_1 < \frac{L_L}{o_0}\right)\frac{1}{L_L}\right] + \frac{1}{o_0+1}\left[\Pr\left(\mathbb{X}_2 > \frac{L_L}{o_0}\right) + \Pr\left(\mathbb{X}_2 > \frac{L_H}{o_0}\right)L_H\right], \quad (8.1)$$

we can express the expected cost above as $c_D + V(o_0)$. As the function V represents the expected cost of making wrong conclusions or no conclusion at all, its value will always be between 0 and 1, as we assume an optimal decision is made, and not making a decision, with cost 1, is always an option.

As the values of L_L, L_H, and o_0 are assumed given, we need only probabilities based on the distributions for $\mathbb{X}_1$ and $\mathbb{X}_2$ in order to compute this cost. In Section 8.2, we will look at methods for estimating such exceedance probabilities with use of simulation.

Example 8.2 Expected cost. In a case one would like to investigate whether two persons are full brothers, or are unrelated. The two hypotheses are denoted H_1 and H_2, respectively. A rather arbitrary prior odds of $o_0 = 1$ is set, and it is planned to analyze DNA samples for the 15 first markers available as `NorwegianFrequencies` in `R Familias`, producing a data set D. The laboratory uses cutoff values $L_H = 10{,}000$ and $L_L = 1/10{,}000$, and we assume the cost of the planned analysis is estimated at $c_D = 0.1$—that is, one is willing to take a cost corresponding to 10 times the cost of the analysis before giving up on finding a solution for the case. As above, we write $\mathbb{X}_1$ and $\mathbb{X}_2$ for the random variables describing LR(D) under H_1 and H_2, respectively. Exercise 8.9 expands on Example 8.10 to show that

$$\Pr(\mathbb{X}_1 < 10{,}000) \approx 0.46,$$

$$\Pr\left(\mathbb{X}_1 < \frac{1}{10{,}000}\right) \approx 1.2 \times 10^{-5},$$

$$\Pr\left(\mathbb{X}_2 > \frac{1}{10{,}000}\right) \approx 0.64,$$

$$\Pr(\mathbb{X}_2 > 10{,}000) \approx 7.9 \times 10^{-6}.$$

Estimating the expected cost with Equation 8.1 gives

$$0.1+0.5 \times \left[0.46+1.2 \times 10^{-5} \times 10{,}000\right] + 0.5 \times \left[0.64 + 7.9 \times 10^{-6} \times 10{,}000\right] = 0.75.$$

As the result is below 1, it seems that the probability of a correct conclusion is high enough to recommend the proposed testing.

Example 8.3 Deciding on obtaining new marker data. As a continuation of the example above, assume the computed posterior odds o based on the actual data turned out to be 100. As $L_L < o < L_H$, no conclusion should be drawn based only on the data D. However, a possibility could be to analyze the existing samples at additional markers, creating new data D'. A reasonable approximation is that D and D' are conditionally independent given hypotheses H_1 and H_2. (This ignores things such as θ corrections explained in Section 2.5 and LD, see Section 4.2, that lead to dependencies between markers, but may be OK in this simplified decision-theoretic context.) We may then use the posterior odds $o = 100$ computed from D as prior odds when determining the value of producing the data D'. Let $\mathbb{X}_1$ and $\mathbb{X}_2$ denote the random variables describing LR(D') under H_1 and H_2, and assume that $c_{D'} = 0.5$. Exercise 8.9 shows that

$$\Pr\left(\mathbb{X}_1 < \frac{L_H}{o_0}\right) = \Pr(\mathbb{X}_1 < 100) \approx 0.65,$$

$$\Pr\left(\mathbb{X}_1 < \frac{L_L}{o_0}\right) = \Pr\left(\mathbb{X}_1 < \frac{1}{1{,}000{,}000}\right) \approx 0.00,$$

$$\Pr\left(\mathbb{X}_2 > \frac{L_L}{o_0}\right) = \Pr\left(\mathbb{X}_2 > \frac{1}{1{,}000{,}000}\right) \approx 1.00,$$

$$\Pr\left(\mathbb{X}_2 > \frac{L_H}{o_0}\right) = \Pr(\mathbb{X}_2 > 100) \approx 9.83 \times 10^{-4}.$$

Equation 8.1 gives an expected cost of

$$0.5 + \frac{100}{100+1}[0.65 + 0.00 \times 10{,}000] + \frac{1}{100+1}\left[1.00 + 9.83 \times 10^{-4} \times 10{,}000\right] = 1.25.$$

As the expected cost is greater than 1, one should not proceed to analyze the extra markers: the cost of the analysis is not justified as it is not likely enough to yield a definite conclusion.

Example 8.4 Genotyping additional relative? Let us return to the situation at the start of Example 8.3. The initial analysis is inconclusive with a posterior odds of $o = 100$. An alternative might be to acquire data D'' from another relative. Such data would, however, not be conditionally independent of D, so we would need to consider

$$\text{LR}(D'' \mid D) = \frac{\Pr(D'' \mid H_1, D)}{\Pr(D'' \mid H_2, D)}.$$

Assume that the cost of the new data is $c_{D''} = 0.1$, and that we estimate, for example based on data from a half-brother, that

$$\Pr\left(\mathbb{X}_1 < \frac{L_H}{o_0}\right) = \Pr(\mathbb{X}_1 < 100) \approx 0.708,$$

$$\Pr\left(\mathbb{X}_1 < \frac{L_L}{o_0}\right) = \Pr\left(\mathbb{X}_1 < \frac{1}{1{,}000{,}000}\right) \approx 0.000,$$

$$\Pr\left(\mathbb{X}_2 > \frac{L_\text{L}}{o_0}\right) = \Pr\left(\mathbb{X}_2 > \frac{1}{1{,}000{,}000}\right) \approx 0.361,$$
$$\Pr\left(\mathbb{X}_2 > \frac{L_\text{H}}{o_0}\right) = \Pr\left(\mathbb{X}_2 > 100\right) \approx 0.001,$$

where $\mathbb{X}_1$ and $\mathbb{X}_2$ now represent the distribution of $\text{LR}(D'' \mid D)$ under H_1 and H_2. Then we get the expected cost of 0.90, and as this cost is less than 1, it is reasonable to go ahead with the data acquisition for the extra relative.

The three examples show the usefulness of studying the distributions of likelihoods if either H_1 or H_2 holds. In particular, the last example shows the usefulness of studying *conditional distributions* of the genotypes of some persons given the genotypes of others, which we will do in Section 8.2.4.

8.1.3 CHOOSING BETWEEN TYPING TECHNOLOGIES

As we saw in Example 8.3, one can use decision theory to decide in a particular case whether one should add additional markers for typing. However, a DNA laboratory will have equipment only for a fixed number of technologies, and the question may be whether one should invest in new equipment to type more markers. This question is generally not related to a specific case. Instead, the question is whether the added ability to make correct decisions in the cases expected by the laboratory with the new equipment warrants its cost.

As an example, assume one plans to initially use standard methods in all cases, but for those cases where the posterior odds is between L_L and L_H, one plans to follow up the analysis by analyzing more markers, using one of several possible technologies. As in the previous section, we make the simplifying assumption that the data from the additional markers is considered independent of the initial marker data given hypotheses H_1 and H_2. If we fix a particular case type (i.e., H_1 and H_2), the cost $c_{D'''}$ of obtaining data D''' with the new technology, and the odds o from the initial data analysis, the cost after the new analysis can be expressed as $c_{D'''} + V(o)$, where the function V is defined in Equation 8.1, and where the variables $\mathbb{X}_1$ and $\mathbb{X}_2$ entering the function V are based on $\text{LR}(D''')$.

As before, one may compare the expected cost with 1 to decide whether one should do the new analysis. One can also investigate $c_{D'''} + V(o)$ as a function of o to find out for which posterior odds a new analysis is worthwhile. Moreover, one may compute V for different technologies and compare them in order to compare their power for resolving cases. Finally, one may also vary the case type over the possible case types expected by the laboratory to see which methods are most suited in which case types.

Note that estimation of V for various values of o depends on evaluating various exceedance probabilities for $\mathbb{X}_1$ and $\mathbb{X}_2$—that is, one will in general need to study the whole distribution.

As making no decision is still an option, with a cost of 1, $V(o)$ will always be at most 1. Generally it will be easier to reach a correct decision when the odds is closer to the limits L_L and L_H, so $V(o)$ will approach zero when o approaches these limits. If two technologies produce data D and D' with similar costs c_D and $c_{D'}$, one may compare their usefulness by comparing their V functions. If one is always smaller than the other, it is universally more powerful in avoiding wrong decisions, and should be preferred. If the two V functions cross each other, their relative utility depends on whether most of the cases one would apply them to have odds values in regions where one or the other is smaller.

Example 8.5 Comparison of typing technologies. In [108] a comparison is made between three different marker sets for their ability to differentiate between various relationships. Three relationship comparisons were investigated:

(a) A standard duo comparison.
(b) A comparison where the alternative to paternity is uncle/nephew.
(c) A comparison where the alternative to paternity is full siblings.

Figure 8.1 shows the function V defined in Equation 8.1 for three different sets of markers: a set of 11 short tandem repeat (STR) markers, a set of 30 deletion and insertion polymorphism (DIP) markers, and a set of 52 single nucleotide polymorphism

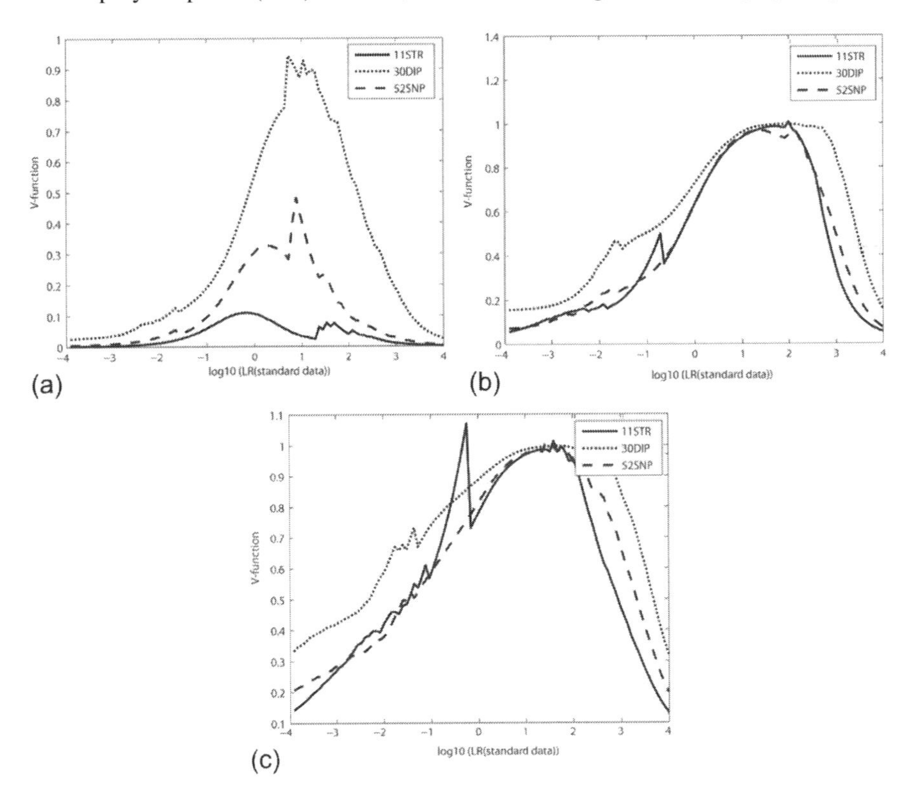

FIGURE 8.1

Examples of V functions comparing different typing technologies.

(SNP) markers. See [108] for further details. Figure 8.1 indicates that the STR marker set is the most powerful in providing information about the comparisons, although the difference from the other marker sets is smaller for relationship comparisons (b) and (c). The jaggedness of the curves Figure 8.1 is the result of simulation errors, as the computations were based on Equations 8.7 and 8.8. Improved simulation techniques, as we will discuss later, can reduce such numerical problems.

8.2 LR AS A RANDOM VARIABLE

In the previous section we defined the two random variables $\mathbb{X}_1$ and $\mathbb{X}_2$ as the ones given by LR(D) when D follows the distribution specified by H_1 or H_2, respectively, and we saw the usefulness of learning about these variables. *Note*: In our context D can take on only finitely many possible values, so both $\mathbb{X}_1$ and $\mathbb{X}_2$ have finite state spaces.

Example 8.6 Mutation in duo case. We start by studying what $\mathbb{X}_1$ and $\mathbb{X}_2$ look like in a very simple case: We consider duo cases with a child and an alleged father:

H_1: The alleged father is the father of the child.
H_2: The alleged father and the child are unrelated.

We use data from a single marker with only two possible alleles A and B, with frequencies p and $q = 1 - p$, respectively. Table 8.1 lists all genotypes that may be observed (i.e., all values of D), their probabilities under H_1 assuming mutations are impossible or assuming mutations happen with probability m, their probabilities under H_2, and the LRs with and without the assumption of mutations. The expressions for the LRs in Table 8.1 follow from Equation 2.12.

From Table 8.1, we see that when we use a model without mutations, LR takes on at most six unique values. When we use a model with mutations, there are generally nine different values.

Table 8.1 Probabilities and LRs for Duo Case and Marker with Two Alleles

AF	CH	H_1 No Mutations	H_1 Mutations	H_2	LR No Mutations	LR Mutations
AA	AA	p^3	$p^3(1-m)$	p^4	$1/p$	$(1-m)/p$
AA	AB	p^2q	$p^2((1-m)q+mp)$	$2p^3q$	$1/(2p)$	$\frac{(1-m)q+mp}{2pq}$
AA	BB	0	p^2mq	p^2q^2	0	m/q
AB	AA	p^2q	p^2q	$2p^3q$	$1/(2p)$	$1/(2p)$
AB	AB	pq	pq	$4p^2q^2$	$1/(4pq)$	$1/(4pq)$
AB	BB	pq^2	pq^2	$2pq^3$	$1/(2q)$	$1/(2q)$
BB	AA	0	pmq^2	p^2q^2	0	m/p
BB	AB	pq^2	$q^2((1-m)p+mq)$	$2pq^3$	$1/(2q)$	$\frac{(1-m)p+mq}{2pq}$
BB	BB	q^3	$q^3(1-m)$	q^4	$1/q$	$(1-m)/q$

Abbreviations: AF, alleged father; CH, child.

The documentation for the function `qkappa` of `BookEKM` provides an example reproducing Table 8.1 and explains how the example can be generalized to any noninbred pairwise relationships, markers having more than two alleles, and arbitrary mutation models.

8.2.1 THE EXPECTATION OF LR

The identity

$$\Pr(\mathbb{X}_1 = x) = x \Pr(\mathbb{X}_2 = x) \tag{8.2}$$

proved in [72] can be used to derive some properties of the distribution of $\mathbb{X}_1$ and $\mathbb{X}_2$ and how they are related. For instance,

$$\Pr(\mathbb{X}_2 > u) = \sum_{x>u} \Pr(\mathbb{X}_2 = x) = \sum_{x>u} \frac{1}{x} \Pr(\mathbb{X}_1 = x) \leq \frac{1}{u}.$$

For $\mathbb{X}_2$ we get

$$\mathrm{E}(\mathbb{X}_2) = \sum x \Pr(\mathbb{X}_2 = x) = \sum \Pr(\mathbb{X}_1 = x) = 1, \tag{8.3}$$

so on an average, the LR will be 1 if the data follows the distribution given by H_2. More generally for $r \geq 1$,

$$\mathrm{E}(\mathbb{X}_2^r) = \sum x^r \Pr(\mathbb{X}_2 = x) = \sum x^{r-1} \Pr(\mathbb{X}_1 = x) = \mathrm{E}\left(\mathbb{X}_1^{r-1}\right),$$

and so, for example,

$$\mathrm{Var}(\mathbb{X}_2) = \mathrm{E}\left(\mathbb{X}_2^2\right) - \mathrm{E}(\mathbb{X}_2)^2 = \mathrm{E}(\mathbb{X}_1) - 1. \tag{8.4}$$

We see that unless the LR is 1 for all D (i.e., $\Pr(D \mid H_1) = \Pr(D \mid H_2)$), the expected value of $\mathbb{X}_1$ is larger than 1. The variance of $\mathbb{X}_2$ determines how much larger.

Example 8.7 RMNE identity. For the most part, our treatment of hypotheses H_1 and H_2 is quite symmetric. However, to make sure $\mathbb{X}_1$ is defined, we have made the assumption that $\Pr(D \mid H_1) > 0$ implies $\Pr(D \mid H_2) > 0$, so $\mathbb{X}_1$ takes on only positive values. However, $\Pr(D \mid H_2) > 0$ does not imply $\Pr(D \mid H_1) > 0$, and $\mathbb{X}_2$ may be zero. In other words, if $\mathbb{E}_i$ is the set of D such that $\Pr(D \mid H_i) > 0$ for $i = 1, 2$, then $\mathbb{E}_1 \subseteq \mathbb{E}_2$, and

$$\begin{aligned} \mathrm{E}\left(\mathbb{X}_1^{-1}\right) &= \sum_{D \in \mathbb{E}_1} \frac{\Pr(D \mid H_2)}{\Pr(D \mid H_1)} \Pr(D \mid H_1) \\ &= \sum_{D \in \mathbb{E}_1} \Pr(D \mid H_2) \\ &= \sum_{D \in \mathbb{E}_2} I\left(\Pr(D \mid H_2) > 0\right) \Pr(D \mid H_2) \\ &= \Pr(\mathbb{X}_2 > 0). \end{aligned} \tag{8.5}$$

Example 8.8 RMNE inequality. In this chapter, we always assume that if data D has a positive probability under H_1, it also has a positive probability under H_2. However, the reverse is not always true. In fact, we may define a *match function*

$$M(D) = I(\Pr(D \mid H_1) > 0)$$

for all data with $\Pr(D \mid H_2) > 0$: if the data has positive probability under H_1 we say we have a "match" and $M(D) = 1$, otherwise we have an "exclusion" and $M(D) = 0$. If we consider $M(D)$ as our data instead of D, we are in the situation of Example 8.1. Note that

$$\begin{aligned} RMNE &= \Pr(M(D) = 1 \mid H_2) \\ &= \Pr(\text{LR}(D) > 0 \mid H_2) \\ &= \Pr(\mathbb{X}_2 > 0). \end{aligned}$$

Combining this with Equation 8.5 we get RMNE $= \text{E}(\mathbb{X}_1^{-1})$. The above may be combined with $\text{E}(\mathbb{X}_1^{-1}) \geq \text{E}(\mathbb{X}_1)^{-1}$, which follows from Jensen's inequality, to give

$$\text{E}(\mathbb{X}_1) \geq \frac{1}{\text{RMNE}}.$$

As remarked in [109], the inverse RMNE can be interpreted as an LR in cases where only information on match or exclusion is retained. By disregarding the complete data, we can expect to underestimate the expected LR when the numerator hypothesis H_1, often referred to as the prosecutor hypothesis, is true. However, this does not mean that it is always conservative to follow the RMNE approach: the inequality applies to expected values and may be reversed for a specific case.

In situations where D is complex it may be tempting to use the simpler variable $M(D)$ for decisions instead of D. However, in general this represents "throwing away information" and may lead to suboptimal decisions. Exercise 8.2 explores this further.

8.2.2 EXPECTATIONS IN SPECIAL CASES

Slooten and Egeland [109] derive a general expression for the expected LR for noninbred pairs of individuals. Recall that these relationships can be described by the parameters $k = (k_0, k_1, k_2)$ denoting the probability that two individuals share zero, one, or two alleles identical by descent (IBD), respectively. Not all specifications of k correspond to a possible pedigree, and Thompson [93] explains why the restriction $k_1^2 \geq 4k_0k_2$ is needed in addition to the obvious $k_0 + k_1 + k_2 = 1$. The expected value of the LR under H_1 for general pairwise relationships can be written

$$E[\mathbb{X}_1] = \alpha L^2 + \beta L + (1 - \alpha - \beta), \tag{8.6}$$

Table 8.2 The Expected LR Under H_1, the Numerator Hypothesis, is shown for Some Noninbred Pedigrees with the Number of Alleles $L = 2, 4, 10$

	k_0	k_1	k_2	$E[\mathbb{X}_1]$	$L=2$	$L=4$	$L=10$
Parent-child	0	1	0	$\frac{L+3}{4}$	1.2500	1.7500	3.2500
Full sibling	$\frac{1}{4}$	$\frac{2}{4}$	$\frac{1}{4}$	$\frac{L(L+7)+24}{32}$	1.3125	2.1250	6.0625
Half sibling	$\frac{1}{2}$	$\frac{1}{2}$	0	$\frac{L+15}{16}$	1.0625	1.1875	1.5625
First cousin	$\frac{3}{4}$	$\frac{1}{4}$	0	$\frac{L+63}{64}$	1.0156	1.0469	1.1406
DFC	$\frac{9}{16}$	$\frac{6}{16}$	$\frac{1}{16}$	$\frac{L^2+31L+480}{512}$	1.0664	1.2109	1.7383
QHFC	$\frac{17}{32}$	$\frac{14}{32}$	$\frac{1}{32}$	$\frac{L^2+127L+1920}{2048}$	1.0635	1.1934	1.6064
Unrelated	1	0	0	1	1.0000	1.0000	1.0000
Monozygous twin	0	0	1	$\frac{L(L+1)}{2}$	3.0000	10.0000	55.0000

Abbreviations: DFC, double first cousin; QHFC, quadruple half first cousin.

where L is the number of alleles,

$$\alpha = \frac{k_2^2}{2},$$

and

$$\beta = \frac{k_1^2 + 4k_1k_2 + 2k_2^2}{4}.$$

Table 8.2 gives examples for some relationships and for markers with 2, 4, and 10 alleles. As $\text{Var}(\mathbb{X}_2) = \text{E}(\mathbb{X}_1) - 1$ according to Equation 8.4, the above result for the expected value under H_1 translates directly to the variance under H_2. Exercise 8.3 confirms the expected value for a parent-child relationship when $L = 2$.

It is rather surprising that the expectation Equation 8.6 depends not on the allele frequencies, but only on the number of alleles L. The variance of $\mathbb{X}_1$ generally depends on the allele frequencies. Some special cases have been dealt with [109]. For instance, when $k_0 = k_1 = 0.5$, corresponding to half sibling, uncle-nephew, and grandfather-grandchild relationships

$$\text{Var}(\mathbb{X}_1) = \frac{1}{256}(-L^2 + 14L - 13) + \frac{1}{128}\sum_{a<b}\left(\frac{p_a}{p_b} + \frac{p_b}{p_a}\right).$$

The minimal value $\frac{1}{256}(L-1)(L+13)$ is attained if and only if all allele frequencies are equal.

8.2.3 ESTIMATING LR EXCEEDANCE PROBABILITIES BY SIMULATION

In Example 8.6 we computed exceedance probabilities for $\mathbb{X}_1$ and $\mathbb{X}_2$ by listing all the possible genotypes, and doing computations for each of them. When the number of possible genotypes becomes large, an alternative is to use simulation: Assume $D_1, D_2, \ldots, D_N$ is a random sample from the data under H_1. Then

$$\Pr(\mathbb{X}_1 < \alpha) \approx \frac{1}{N} \sum_{i=1}^{N} I(\mathrm{LR}(D_i) < \alpha), \tag{8.7}$$

where $I()$ is the indicator function, and is 1 if and only if its contents are true. Similarly, if $D'_1, D'_2, \ldots, D'_N$ is a random sample from the data if H_2 holds, then

$$\Pr(\mathbb{X}_2 > \alpha) \approx \frac{1}{N} \sum_{i=1}^{N} I(\mathrm{LR}(D'_i) > \alpha). \tag{8.8}$$

The simulation method of Equations 8.7 and 8.8 is simple and direct to implement. However, to optimize the accuracy in the estimation of a particular exceedance probability, it is better to make sure many values are simulated in the region close to the cutoff value α, while at the same time making adjustments for such a skewed simulation. Importance sampling is a general method that achieves this, as explained in the present context in [111]. For example, we may write

$$\begin{aligned}
\Pr(\mathbb{X}_2 > \alpha) &= \sum_{D} I(\mathrm{LR}(D) > \alpha) \Pr(D \mid H_2) \qquad (8.9)\\
&= \sum_{D} I(\mathrm{LR}(D) > \alpha) \times \frac{\Pr(D \mid H_2)}{\Pr(D \mid H_1)} \times \Pr(D \mid H_1)\\
&= \sum_{D} \frac{I(\mathrm{LR}(D) > \alpha)}{\mathrm{LR}(D)} \Pr(D \mid H_1)\\
&\approx \frac{1}{N} \sum_{i=1}^{N} \frac{I(\mathrm{LR}(D_i) > \alpha)}{\mathrm{LR}(D_i)},
\end{aligned}$$

where as before $D_1, D_2, \ldots, D_N$ is a random sample for the data under assumption H_1. If α is such that values of $\mathrm{LR}(D)$ are likelier to be close to α when D follows the distribution of H_1 than when D follows the distribution of H_2, using Equation 8.9 is likely to improve accuracy. The simulation is exemplified below. A simple example with one marker is explored in detail and this is followed by an example dealing with the calculations in Examples 8.2 and 8.3.

Example 8.9 Importance sampling with once marker. We illustrate importance sampling for a simple case with one SNP marker with allele frequencies p and $1 - p$. There are two individuals and the hypotheses are "brothers" (H_1) and "unrelated" (H_2). Table 8.3 lists the possible values for $x = \mathrm{LR}$ when $p = 0.6$. For instance the maximal value 3.0625 occurs with probability $(1 - p)^4 = 0.02560$ when H_2 is true. The `tab` produced in the `R` code below corresponds to Table 8.3.

```
require(DNAprofiles)
p <- 0.6
db <- list(c(p,1-p))
```

Table 8.3 The Distribution of the LR for a SNP Marker with Allele Frequencies 0.6 and 0.4 for Two Individuals if They are Brothers (Middle Column) or Unrelated (Rightmost Column)

	x	$\Pr(\mathbb{X}_1 = x)$	$\Pr(\mathbb{X}_2 = x)$
1	0.25000	0.02880	0.11520
2	0.66667	0.23040	0.34560
3	0.87500	0.13440	0.15360
4	1.29167	0.29760	0.23040
5	1.77778	0.23040	0.12960
6	3.06250	0.07840	0.02560

```
names(db) <- "Marker1"
# The LR distributions are calculated next using
# the function ki.dist in DNAprofiles
h1 <- ki.dist(hyp.1 = "FS", hyp.2 = "UN", hyp.true = "FS",
              freqs.ki = db)
h2 <- ki.dist(hyp.1 = "FS", hyp.2 = "UN", hyp.true = "UN",
              freqs.ki = db)
tab <- data.frame(h1,h2)[,-3]
```

In our case, all calculations of exceedance probabilities can be done without simulation. For instance,

$$\Pr(\mathbb{X}_2 > 1) = 0.23040 + 0.12960 + 0.02560 = 0.3856.$$

Direct Monte Carlo simulations, for example with

```
N <- 10^6
LR <- sample(tab[, 1], N, prob = tab[, 3], rep = T)
length(LR[LR>1])/N
```

will also give an accurate answer for this relatively large probability. Consider next how the probability is estimated with the use of importance sampling as summarized by Equation 8.9. First, generate $N = 10^6$ samples from the distribution of $\mathbb{X}_1$, with

```
set.seed(99999)
LR <- sample(tab[, 1], 1e+06, prob = tab[,2], rep = T)
I <- as.integer(LR > 1)
```

The first 10 samples are shown in Table 8.4. The importance sample estimate is obtained by `mean(I/LR)`, which according to Equation 8.9 gives

$$\Pr(\mathbb{X}_2 > 1) \approx \frac{1}{10^6}\left(\frac{1}{1.77778} + \frac{0}{0.87500} + \cdots + \frac{1}{0.6666667}\right) = 0.3857.$$

Table 8.4 Ten Simulated LR Values are shown Along with the Indicator which is 1 if LR Exceeds 1 and is 0 Otherwise

	LR	I
1	1.77778	1
2	0.87500	0
3	1.77778	1
4	0.87500	0
5	1.29167	1
6	1.29167	1
7	0.87500	0
8	1.77778	1
9	0.25000	0
10	1.29167	1

Notes: *These are obtained under the assumption the individuals are brothers according to the probabilities in Table 8.3.*

A slight modification of the example will demonstrate the potential usefulness of importance sampling: Let $p = 0.01$. Then the maximal LR $(0.25p^2 + 0.5p + 0.25)/p^2 = 2550.25$ occurs with probability $p^4 = 0.00000001$ when H_2 is true. In our one-marker example, direct calculations show that, $\Pr(\mathbb{X}_2 > 1000) = 0.00000001$, but this will not be estimated accurately by direct Monte Carlo simulation. For instance, with $N = 10^6$ simulations the probability is

$$\left(1 - p^4\right)^N = 0.99$$

that the estimate will be 0. Modification of the above R code results in the accuracy of importance sampling being verified.

Example 8.10 Probabilities in LRs. This example builds on Example 8.2 and demonstrates how to obtain exceedance probabilities like those used there to calculate the expected cost. The below code calculates the distribution of $\mathbb{X}_1$ and $\mathbb{X}_2$ for the 15 first markers in the database:

```
require(Familias); require(DNAprofiles)
data(NorwegianFrequencies)
set.seed(17)
h1 <- ki.dist(hyp.1 = "FS", hyp.2 = "UN", hyp.true = "FS",
              freqs.ki = NorwegianFrequencies[1:15])
h2 <- ki.dist(hyp.1 = "FS", hyp.2 = "UN", hyp.true = "UN",
              freqs.ki = NorwegianFrequencies[1:15])
```

Relationships other than full siblings (abbreviated "FS" above) can be specified as can be seen from the documentation for the function `ki.dist` in `DNAprofiles`. To estimate $\Pr(\mathbb{X}_1 < 10{,}000)$ direct simulation suffices, and the result 0.46 is obtained by entering

```
set.seed(17)
1-sim.q(t=10000, N=1e6, dists=h1)
```

The above function `sim.q` of `DNAprofiles` also accommodates importance sampling. To estimate $\Pr(\mathbb{X}_2 > 10{,}000)$, we sample from the distribution under H_1 as then larger LRs are likelier to occur. The code below yields the result 7.9×10^{-6}:

```
sim.q(t=10000, N=1e6, dists=h2, dists.sample = h1)
```

Importance sampling will generally give a more accurate estimate when we simulate from a distribution that is more concentrated in the region of interest than the original distribution. An indication of this in the example above is that the variability is much smaller for importance sampling. Further calculations for Examples 8.2 and 8.3 are explored in Exercise 8.9.

8.2.4 CONDITIONAL SIMULATION

Simulation was briefly introduced in Section 2.10 for `Familias` and exemplified in Exercise 2.17. The programs `FamLink` and `FamLinkX` presented in Section 4.4 also implement simulation. These simulations are *unconditional*: it is not possible to sample genotype data given observations for other individuals. Below we focus on situations when some individuals have been genotyped. This information needs to be accounted for, and conditional simulation is called for. If the genotyped individuals are founders—that is, their parents are not in the pedigree, the simulation algorithm intuitively described as "allele dropping" is easily implemented. Below we describe conditional simulation and implementation in `R` for gradually more complicated cases.

Example 8.11 Conditional simulation. `DNAprofiles`. The `R` package `DNAprofiles` provides functions to do conditional simulation for pairs of noninbred individuals. The pedigree can be specified by an abbreviation, such as "FS" for full siblings, or more generally by the three IBD parameters as explained in Section 8.2.2. Below, first a random profile `x` based on the allele frequencies for the marker `SE33` of the database `NorwegianFrequencies` is sampled and then three profiles for a full sibling of `x` are simulated:

```
require(Familias); require(DNAprofiles)
data(NorwegianFrequencies)
set.seed(123)
x <- sample.profiles(1, freqs = NorwegianFrequencies["SE33"])
x.FS <- sample.relatives(x, 3, type = c(0.25,0.5,0.25))
res.index <- rbind(x,x.FS)
```

```
L1 <- names(NorwegianFrequencies[["SE33"]])
data.frame(apply(res.index,2, function(x) L1[x]))
```

```
  SE33.1 SE33.2
1     19     14
2     19     12
3   27.2   21.2
4     17     20
```

The above functions work with the index of the allele frequency vector. The last line of code produces output with the original names for the alleles. The second line of the output is a simulated genotype of the full sibling of the individual in line 1, and similarly for lines 3 and 4. Note that these four individuals are not full siblings. For one thing more than four alleles appear above; this is a pairwise simulation. LRs can be calculated with use of the function `ki` as exemplified in Exercise 8.7.

There are methods and implementations for general conditional simulation. The program `SLINK` of the `LINKAGE/FASTLINK` suite handles conditional simulation.[2] We next demonstrate a modern `R` implementation (with various time savings compared with `SLINK`)—namely, `markerSim` of the package `paramlink`. This latter implementation also works for inbred pedigrees.

Example 8.12 Conditional simulation, `paramlink`. The above example is continued, but now we simulate genotypes of three full siblings given a specific genotype of a fourth full sibling. See the help pages of `paramlink` for an explanation.

```
require(paramlink)
x <- nuclearPed(4, sex=1)
L1 <- names(NorwegianFrequencies[["SE33"]])
freq <- as.double(NorwegianFrequencies[["SE33"]])
m <- marker(x, alleles=L1, afreq=freq, 3, c(14,19))
simPed <- markerSim(x, N=5, available = c(4,5,6),
                    partialmarker = m, seed = 17, verbose=FALSE)
```

The pedigree with the five simulated sets of genotypes is plotted by entering

```
plot(simPed, marker=1:5)
```

In the next two examples we describe the calculations required for Example 8.4. Recall that exceedance probabilities for

$$\mathrm{LR}(D'' \mid D) = \frac{\Pr(D'' \mid H_1, D)}{\Pr(D'' \mid H_2, D)} \tag{8.10}$$

are needed to calculate the expected cost.

Example 8.13 Data from more relatives – Part I. In a case it is investigated whether two men are full brothers or unrelated. To help settle the case, obtaining data from a known half-brother of one of the men is considered. Specifically, under H_1,

[2]See http://linkage.rockefeller.edu/ott/SLINK.htm.

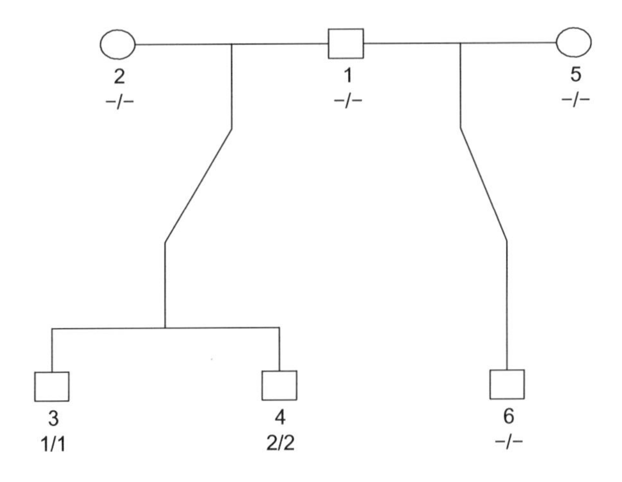

FIGURE 8.2

Illustration of the pedigree for Example 8.13.

the men are related as individuals 3, 4, and 6 in Figure 8.2, whereas under H_2, 3 and 6 are related as in Figure 8.2, whereas 4 is unrelated to them.

The data already available, D, is shown in Figure 8.2. We are considering obtaining new data D'' from individual 6. Assume first we are dealing with a SNP marker with allele frequencies $p = 0.7$ and $q = 1 - p$. Table 8.5 shows the LRs for the possible genotypes individual 6 in Figure 8.2 may have—that is, 1/1, 1/2, and 2/2 and their probabilities under H_1 and H_2. For example, under H_1, the parents 1 and 2 must both be 1/2. Therefore, the conditional probability that individual 6 is 1/1 is $p/2 = 0.35$. Under H_2, this probability is $p/2 + p^2/2 = 0.595$, and therefore $\text{LR} = 0.35/0.595 = 0.58824$ if individual 6 is 1/1. If there were more than two alleles and mutation and other artifacts were ignored, exclusion would occur under H_1 for, say, individual 6 being 3/3.

Example 8.14 Data from more relatives – Part II. This example uses the same hypotheses as Example 8.13, but more realistic data. We now assume data is available for individuals 3 and 4 in Figure 8.2 for 15 markers. The conditional distributions for the possible genotypes of individual 6 under H_1 and H_2 are obtained with use of the

Table 8.5 The Probability Distribution in Example 8.13

	LR	$\Pr(\mathbb{X}_1)$	$\Pr(\mathbb{X}_2)$
1/1	0.58824	0.35000	0.59500
1/2	1.38889	0.50000	0.36000
2/2	3.33333	0.15000	0.04500

function `oneMarkerDistribution` of `paramlink`. From these probabilities, the LR distribution can be obtained for each marker and each hypothesis. The function `sim.q` of `DNAprofiles` can then be used to estimate the required probabilities. The complete code is included as an example in `BookEKM`.

8.3 EXERCISES

Exercise 8.1 Decision procedure in paternity cases. Assume a DNA laboratory uses the following decision procedure in its standard paternity cases. We compute

$$\mathrm{LR} = \frac{\Pr(\text{data} \mid H_1\text{: paternity})}{\Pr(\text{data} \mid H_2\text{: no paternity})}$$

and compare it with cutoff values C_H and C_L. If $\mathrm{LR} > C_\mathrm{H}$, one concludes for paternity, and if $\mathrm{LR} < C_\mathrm{L}$, one concludes against paternity, while in other cases there is no conclusion (which may lead to more data being acquired). In standard paternity cases, the laboratory believes the prior odds for paternity can be specified with a fixed number $o_{0,\text{standard}}$.

(a) The laboratory believes the constants C_H and C_L chosen represent an optimal balance between various risks. Assuming this is correct, show that the implied costs c_1 and c_2 can be expressed as

$$c_1 = 1/(o_{0,\text{standard}} C_\mathrm{L}),$$
$$c_2 = o_{0,\text{standard}} C_\mathrm{H}.$$

(b) Now, assume the laboratory receives a special paternity case where the prior odds o_0 is assumed to be smaller than the standard prior odds $o_{0,\text{standard}}$. Show that in this case there exist numbers x which satisfy $x > C_\mathrm{H}$ but also

$$\frac{1}{c_1} < o_0 \times x < c_2.$$

(c) Assume an LR satisfying the conditions above for x has been computed. Then the laboratory would conclude for paternity, whereas the optimal decision would be no conclusion. Compute a formula for the expected cost of concluding for paternity, and show that this value is always larger than 1. The laboratory's decision in this case is suboptimal compared with making no decision, which has a cost 1.

Exercise 8.2 Decision procedure in paternity cases. RMNE. Assume a DNA laboratory is facing a duo case, with data D consisting of DNA samples from a man S and his alleged father, and with the following two alternative hypotheses:

H_1: The alleged father is the biological father.
H_2: A randomly sampled man from the relevant population is the biological father.

Assume a prior odds of $o_0 = 1$. To do the computations, a probability model is used where some data that may occur under H_2 is impossible under H_1 (e.g., one uses a model without mutations). To simplify matters, the laboratory considers the option of basing its conclusion on whether a match (i.e., $\Pr(D \mid H_1) > 0$) or an exclusion (i.e., $\Pr(D \mid H_1) = 0$) has been found. Assume the laboratory generally uses likelihood cutoffs L_H and L_L, thereby indirectly specifying its costs for making erroneous decisions. Reformulating the exclusion decision rule in terms of the LR, the following table summarizes the decision rules the laboratory considers (see Example 8.1):

	LR = 0	$0 < \text{LR} < L_L$	$L_L \leq \text{LR} \leq L_H$	$L_H < \text{LR}$
Standard rule	H_2	H_2	undecided	H_1
Exclusion rule (RMNE $< o_0/2$)	H_2	H_1	H_1	H_1
Exclusion rule (RMNE $\geq o_0/c_2$)	H_2	undecided	undecided	undecided

(a) Assume the laboratory performs the DNA analysis and obtains a specific LR. Show that the expected cost of making the decision H_1 given this value of LR can be expressed as

$$\frac{1 + L_H}{1 + \text{LR}}$$

and that the expected cost of making the decision H_2 given LR can be expressed as

$$\frac{\text{LR}/L_L + \text{LR}}{1 + \text{LR}}.$$

(b) For each of the three cases $0 < \text{LR} < L_L$, $L_L \leq \text{LR} \leq L_H$ and $L_H < \text{LR}$ use the formulas above to find which of the three possible decisions (deciding on H_1, deciding on H_2, or making no decision, which has cost 1) has the lowest expected cost, and compute in terms of LR the expected increase in cost incurred by making one of the suboptimal decisions.

Exercise 8.3 *LR as a random variable. `paramlink`. The LR can in some circumstances be treated as a random variable as discussed in Section 8.2 and exemplified below.

Consider a single diallelic marker. The alleles are denoted a and b, and the frequencies are p and $1 - p$. For the numerical examples, $p = 0.5$. In this case we let H_1 state that there is a parent-child relationship between the alleged father and the child, and let H_2 state that they are unrelated. Throughout we make the standard simplifying assumptions of Hardy-Weinberg equilibrium, no mutations, no silent alleles, and no genotyping error.

(a) Show that

$$\Pr(G_{AF} = a/a, G_{CH} = b/b \mid H_2) = p^2(1-p)^2 = 0.0625.$$

Note: The probability of the same event under H_1 is 0.

(b) Use `paramlink` to obtain the complete distribution of G_{AF} and G_{CH} under H_1 and H_2 given in Tables 8.6 and 8.7.

(c) Find $\Pr(\mathbb{X}_1 = 0)$ and $\Pr(\mathbb{X}_2 = 0)$.

(d) Reproduce Table 8.8.

(e) Verify Equation 8.5

$$E(\mathbb{X}_1^{-1}) = \Pr(\mathbb{X}_2 > 0)$$

on the basis of Table 8.8.

Exercise 8.4 Expectation of LR. The previous exercise is continued by our considering expected values of the LR.

(a) Show that $E[\mathbb{X}_1] = 1.25$ and $E[\mathbb{X}_2] = 1$ on the basis of Table 8.8. Calculate $E[\mathbb{X}_1]$ once more, this time with use of Equation 8.6.

(b) Consider two pedigrees, Ped_1 and Ped_2, specifying noninbred pairwise relationships. Assume that for Ped_1 there is a positive probability that the individuals share two alleles IBD, while this probability is 0 for Ped_2. Explain why the expected LR for ped1 is greater than that for Ped_2 for a marker containing a sufficient number of alleles (in both cases the numerator of LR specifies the individuals are unrelated).

Table 8.6 The Distribution of G_{AF} (Rows) and G_{CH} Under H_1 Requested in Exercise 8.3

	a/a	*b/b*	*a/b*
a/a	0.1250	0.0000	0.1250
b/b	0.0000	0.1250	0.1250
a/b	0.1250	0.1250	0.2500

Table 8.7 The Joint Distribution of G_{AF} (rows) and G_{CH} Under H_2 Requested in Exercise 8.3

	a/a	*b/b*	*a/b*
a/a	0.0625	0.0625	0.1250
b/b	0.0625	0.0625	0.1250
a/b	0.1250	0.1250	0.2500

Table 8.8 The Distribution of LR Under H_2 and H_1 Requested in Exercise 8.3

x	$\Pr(\mathbb{X}_1 = x)$	$\Pr(\mathbb{X}_2 = x)$
0	0	0.125
1	0.75	0.75
2	0.25	0.125

Exercise 8.5 *Exclusion power. Simulation. paramlink. This exercise deals with the power of exclusion (PE) as implemented in the paramlink function exclusionPower. The methods and implementation are described in [49], and we briefly review the notation. It suffices to consider one marker as independence is assumed. There are N individuals with a claimed relationship described by a pedigree Ped_1. Typically there will be more than N individuals in Ped_1 as additional persons may be needed to define the family relationships. We assume that all N will be available for genotyping, but we allow for situations where some genotypes are known from previous genotyping (all analyses are then conditional on these). If the true relationship is described by Ped_2, the marker's PE is defined as

$$\text{PE} = 1 - \text{RMNE} = \Pr(\text{Ped}_1 \text{ incompatible with genotypes} \mid \text{Ped}_2). \tag{8.11}$$

(a) Consider first a standard paternity case and a SNP marker with allele frequencies p and $1-p$. No individuals are genotyped, but genotype data will eventually be available from the alleged father and the child. Show that $\text{PE} = 2p^2(1-p)^2$. Check the formula for $p = 0.2$ by entering

```
require(paramlink)
claim <- nuclearPed(noffs = 1, sex = 2)
true <- list(singleton(id = 1, sex = 1), singleton(id = 3,
                  sex = 2))
available <- c(1, 3)
exclusionPower(claim, true, available,
                    alleles = 2, afreq = c(0.2, 1-0.2))
```

Observe that a plot is produced by default.

(b) Assume the child is known to have genotype 1/1. Calculate PE for this case.

(c) Do problem (a) above when the marker is on the X chromosome.

(d) Repeat the above exercises for a marker with allele frequencies 0.7, 0.1, 0.1, and 0.1.

Next, consider a more complicated example: two females claim to be mother and daughter. Below we compute the power to reject this claim if in reality they are sisters for a SNP marker.

(e) Define the pedigrees by typing

```
mother.daughter <- nuclearPed(1, sex = 2)
sisters <- relabel(nuclearPed(2, sex = c(2, 2)),
                   c(101, 102, 2, 3))
```

Calculate the probability of exclusion for an equifrequent SNP marker.

(f) Consider next an extreme case of inbreeding corresponding to the parents being full siblings. Calculate the power of exclusion by entering

```
sisters.LOOP <- addParents(sisters, 101, father = 201,
                           mother = 202)
sisters.LOOP <- addParents(sisters.LOOP, 102, father = 201,
                           mother = 203)
exclusionPower(ped_claim = mother.daughter,
               ped_true = sisters.LOOP, loop = 101,
               ids = c(2, 3), alleles = 2)
```

Assume the minor allele frequency is 0.1 and that one of the sisters—say, 3—is homozygous for the minor allele. Redo the last calculation.

Exercise 8.6 Conditional simulation paramlink.

(a) The function `markerSim` of `paramlink` can be used to simulate marker data as explained in Example 8.12. In this exercise we consider two genotyped brothers and we wish to simulate genotypes for their parents. Enter the following commands to perform five simulations based on the marker `SE33` from the database `NorwegianFrequencies` under the assumption that the brothers are 11/12 and 12/13:

```
x <- nuclearPed(2, sex=1)
require(Familias);
data(NorwegianFrequencies)
L1 <- NorwegianFrequencies[["SE33"]]
m <- marker(x, alleles=names(L1), afreq=L1,
            3, c(11,12), 4, c(12,13))
simPed <- markerSim(x, N=5, available = c(1,2),
          partialmarker = m, seed = 17,verbose=FALSE)
```

(b) Plot the pedigree with genotypes indicated.

Exercise 8.7 Simulation. LR. `DNAprofiles`. Some functionality of `DNAprofiles` involving simulation and calculation of LRs for pairwise relationships was introduced in Example 8.11. This is expanded on below.

(a) Load the package, the database `freqsNLngm`, and sample a random DNA profile `x`, by entering

```
set.seed(123)
require(DNAprofiles)
```

```
data(freqsNLngm)
x <- sample.profiles(N = 1, freqs = freqsNLngm)
```

Note: A summary of the database is obtained by, for example, `str(freqsNLngm)`.

(b) Sample a DNA profile of a full sibling of the individual with profile `x` by typing

```
set.seed(123)
x.FS <- sample.relatives(x, 1, type = "FS")
```

Consider hypothesis H_1 stating that individuals with profiles x and x.FS are full siblings versus hypothesis H_2 stating unrelatedness. Calculate the full sibling index (i.e., the LR) with use of the function `ki`.

(c) Sample DNA profiles for 10,000 full siblings of the individual with profile `x`. Calculate all sibling indices. Find the smallest and the largest sibling index. Comment on the answer. Make a histogram of $\log_{10}(\text{SI})$, where SI is the sibling index. Estimate the probability that LR is below 1.

Exercise 8.8 *Importance sampling. `DNAprofiles`. In this exercise we will illustrate *importance sampling*, which was introduced in Section 8.2.3, first generally, and then using the package `DNAprofiles`.

(a) Let $U \sim N(\mu, 1)$. Find $\Pr(U > 5 \mid \mu = 0)$ with use of `pnorm`.

(b) Note that $q = \Pr(U > 5 \mid \mu = 0) = E(I(U > 5) \mid \mu = 0)$ can, in principle, be estimated by Monte Carlo simulation. Generate $N = 10^6$ samples and estimate q.

(c) Direct Monte Carlo simulation, as above, works poorly when it comes to estimating small probabilities. An alternative is importance sampling, which in this case can be achieved by writing

$$q = E(I(U > 5) \mid \mu = 0) = E\left(I(U > 5)\frac{\phi_0(U)}{\phi_5(U)} \mid \mu = 5\right),$$

where ϕ_μ is the probability density function of U. Estimate q by importance sampling.

(d) For the remainder of this exercise we will estimate exceedance probabilities $q_{t|H} = \Pr(\text{LR} > t \mid H)$. The estimator

$$\tilde{q}_t \mid H_2 = \frac{1}{N}\sum_{i=1}^{N} I(Z_i > t)w(Z_i) \tag{8.12}$$

was introduced in [111] in the present context. Here Z_i are LRs simulated from H_1 and $w(Z_i) = 1/Z_i$ as shown in [112]. Run the example in the documentation for `sim.q`—that is,

```
require(DNAprofiles)
data(freqsNLngm)
hp <- ki.dist(hyp.1 = "PO", hyp.2 = "UN", hyp.true = "PO",
```

```
                    freqs.ki = freqsNLngm)
hd <- ki.dist(hyp.1 = "PO", hyp.2 = "UN", hyp.true = "UN",
                    freqs.ki = freqsNLngm)
set.seed(100); N <- 1e5; t <- 1e6
q <- sim.q(t = t, dists = hd, dists.sample = hp, N = N)
```

(e) Let $n = 15$ denote the number of markers. Note that

$$q_{0|H} = \Pr(\mathrm{LR} > 0 \mid H) = \prod_{i=1}^{n} \Pr(\mathrm{LR}_i > 0 \mid H)$$
$$= \prod_{i=1}^{n} (1 - \Pr(\mathrm{LR}_i = 0 \mid H)).$$

Estimate $q_{0|H_\mathrm{d}}$ and compare it with the exact value. *Hint and comment*: For $t = 0$ (or for $t < 0$, which gives an estimate of $\Pr(\mathrm{LR} \geq 0 \mid H_\mathrm{d})$) it is preferable to sample from H_d and not use the biasing distribution H_p. An exact alternative is to use the above equation directly.

(f) Next only the first marker D1S1656 is considered—that is, `freqsNLngm[1]`. Let t be the median of LR outcomes under H_d. Calculate $q_{t|H_\mathrm{d}}$ exactly and compare it with the simulated counterpart based on $N = 10{,}000$. Simulate 100 q estimates and plot a density estimate.

Exercise 8.9 Importance sampling (see Examples 8.2 and 8.3).

(a) Verify the calculations in Example 8.2 by expanding on Example 8.10.

(b) Estimate $\Pr(\mathbb{X}_2 > 10{,}000)$ ten times with and without importance sampling. Comment on the results.

(c) Do the calculations for Example 8.3.

Glossary for non-biologists

Below is a list of terms and concepts connected to the biology of forensic markers and the technology of their measurements, together with a short definition or discussion of each term. Some concepts are explained using yet other concepts that themselves appear in the glossary. These are also written in *italics* to indicate that a proper explanation exists elsewhere in the glossary.

Allele Different variants at a *marker*, see also *short tandem repeat* and *single nucleotide polymorphism*.

Allelic association See *linkage disequilibrium*.

Autosomal Term used to refer to the DNA residing on the chromosomes that are not sex specific. Autosomal *chromosomes* are typically referred to using integers 1-22, while sex chromosomes are denoted as either X or Y.

Capillary electrophoresis The separation of genetic fragments, amplified in a *polymerase chain reaction*, resulting in the genetic *profile*. Visualized in an *electropherogram*.

Chromosome Human DNA is divided into 46 different subsets, known as chromosomes. Some independence can usually be assumed between *markers* at different chromosomes—for example, *linkage* is considered to only stretch the length of an individual chromosome.

Coancestry A measure of the degree of common ancestry.

Degraded DNA Defined here as DNA of low quality. Usually observed in a genetic *profile* as a "ski slope" effect, where short *STR markers* tend to be less affected by degradation than longer markers. For degraded DNA profiles, observational level errors such as *dropouts* and *drop-ins* are commoner.

Drop-in An observation-level effect that may occur in a genetic *profile* and causes an *allele* to appear in a profile. For instance, a true *homozygous* profile may be genotyped as a *heterozygote* owing to the drop-in of an extra allele.

Dropout An observation-level effect that may occur in a genetic *profile* caused when the *polymerase chain reaction* fails to completely amplify the *genotype*. In other words, one of the alleles are not observed. The phenomenon may occur in a *homozygous* or *heterozygous* genotype, but is, under normal circumstances, only visible in the latter case.

Electropherogram Illustration of the genetic *profile* of an individual following *capillary electrophoresis*. See also *polymerase chain reaction*.

Gamete In humans the gametes include the female eggs and the male sperms. Each gamete contains only half of the individual's *genome*, chosen in a process known as *meiosis*.

Genome The complete genetic setup of an individual. For humans this encompasses 23 pairs of *chromosomes*, on average 3.2×10^9 base pairs.

Genotype An individual inherits one allele/variant/gene from his/her mother and father. The joint contribution is called the genotype. See also *homozygous* and *heterozygous*.

Germ line See *mutation*.

Haplotype Combination of *alleles* at different *markers*. A haplotype indicates alleles located on the same *chromosome*, in contrast to a *genotype*, which indicates combination of alleles on different chromosomes. See also *Phased genotype* and *Unphased genotype*.

Hardy-Weinberg equilibrium Common assumption when computing *genotype* probabilities. Assumes that the *allele* frequency distribution is at equilibrium and is not affected by effects such as nonrandom mating, population substructure, *mutations*, and natural selection.

Heterozygote A *genotype* consisting of two different *alleles*—for example, 12/13 or A/a.

Homozygote A *genotype* consisting of two identical *alleles*—for example, 12/12 or A/A.

Identical by descent Two alleles originating from the same ancestral *allele* are identical by descent. Two alleles in one individual may be identical by descent.

Identical by state Two *alleles* identical by appearance are identical by state. Two alleles may be identical by state but not necessarily *identical by descent*, while the opposite must be true if we do not consider mutations.

Linkage A pedigree-level effect causing adjacent *markers* to be inherited as a unit within a pedigree. Measured in centimorgans (cM). Not to be confused with *linkage disequilibrium*.

Linkage disequilibrium A population level effect caused by the accumulation of some *haplotypes* in the population. The effect is measured as the deviation between expected and observed haplotype frequencies.

Locus A particular position in a *genome* (plural, loci). See also *marker*.

Marker Location/position on the chromosomes where different *alleles* are observed. Markers may be of different types; see *single nucleotide polymorphism* and *short tandem repeat*.

Maternal Term used to refer to something that originates from the mother of an individual. For instance, maternal *allele* refers to the allele inherited from the mother.

Meiosis A process occurring in the human cells creating the *gametes* (eggs and sperms). Also known as the division of sex cells.

Microarray Technology where the DNA is analyzed on a small chip, typically high-density *single nucleotide polymorphism* arrays.

Mitochondrial DNA Term used to refer to DNA located outside the nucleus of the cells, more specifically the mitochondrias. Inherited from mother to all her biological children. Commonly abbreviated mtDNA.

Mutation A change in the DNA sequence initiated by a random or accidental event. A mutation may be somatic, where it is inherited only when the cell divides within the individual, or in a sex cell (germ line), where it is transferred to the next generation. In this book we are concerned with the latter type of mutation.

Mutation rate The rate at which a *mutation* is expected to occur. For *short tandem repeat* markers the rate is high, roughly 0.5%, while for *single nucleotide polymorphism* markers the rate is low, typically below 10^{-8}.

Next-generation sequencing Refers to second-generation sequencing and encompasses several different techniques used to obtain the specific genomic sequence of an individual.

Paternal Term used to refer to something that originates from the father of an individual. For instance, paternal *allele* refers to the allele inherited from the father.

PCR Polymerase chain reaction is the process where the DNA is amplified into virtually unlimited numbers. More specifically, individual *markers* may be selectively amplified. See also *capillary electrophoresis*.

Phased genotype A genotype that includes information about whether alleles are inherited from the father or the mother.

Phenotype The observed physical appearance of some genetic trait. The connection between the observed *genotypes* and phenotypes can be modeled in calculations of likelihoods. However, for the scope of this book, phenotypes are not accounted for.

Polymorphism In genetics defined as the degree of variation at a *marker*, or more specifically the number of different *alleles* and the distribution of their frequencies.

Profile The *genotypes* for a set of genetic *markers* for an individual.

SNP Single nucleotide polymorphism (often pronounced "snip"). A type of genetic *marker* where a single nucleotide base has been substituted at a specific location on the *chromosome*. Potentially four different *alleles* may exist at each single nucleotide polymorphism: A, C, T, or G. In practice, two, or more seldom three, different alleles are observed. Their number in the *genome* is expected to exceed 80,000,000.

STR Short tandem repeat. A subset of genetic *markers* where the *alleles* are defined as a number indicating the number of repeats—for example, the sequence `AACG` may be repeated 12 times. Different alleles, therefore, have different lengths. Currently the preferred marker choice in forensic genetics, partly owing to their high *polymorphism*.

Trio Standard relationship case, commonly consisting of a child and the undisputed mother and an alleged father.

Trisomy The case where the genotype of an individual consists of three *alleles*, a consequence of malfunction during the *meiosis* leading to the transmission of two alleles from one of the parents.

Typing kits Refers to sets of genetic *markers* amplified in a single *polymerase chain reaction*. Several commercial typing kits exist, including PowerPlex ESX17 (Promega) and GlobalFiler (Life Technologies).

Unphased genotype A genotype that does not include information about whether alleles are inherited from the father or the mother. Genotype usually implies unphased genotype.

Bibliography

[1] J. Bowers, J.-M. Boursiquot, P. This, K. Chu, H. Johansson, C. Meredith, Historical genetics: the parentage of Chardonnay, Gamay and other wine grapes of northeastern France, Science 285 (1999) 1562-1565.

[2] T. Egeland, P. Mostad, B. Olaisen, A computerised method for calculating the probability of pedigrees from genetic data, Science & Justice 37 (4) (1997) 269-274.

[3] T. Egeland, P. Mostad, B. Mevåg, M. Stenersen, Beyond traditional paternity and identification cases. Selecting the most probable pedigree, Forensic Sci. Int. 110 (2000) 47-59.

[4] T. Egeland, P. Mostad, Statistical genetics and genetical statistics: a forensic perspective, Scand. J. Stat. 29 (2002) 297-307.

[5] D. Kling, T. Egeland, A. Tillmar, FamLink—a user friendly software for linkage calculations in family genetics, Forensic Sci. Int. Genet. 6 (5) (2012) 616-620.

[6] D. Kling, B. Dell'Amico, A.O. Tillmar, FamLinkX—implementation of a general model for likelihood computations for X-chromosomal marker data, Forensic Sci. Int. Genet. 17 (2015) 1-7.

[7] D. Kling, B. Dell'Amico, P.J.T. Haddeland, A.O. Tillmar, Population genetic analysis of 12 X-STRs in a Somali population sample, Forensic Sci. Int. Genet. 11 (2014) e7-e8.

[8] D. Kling, J. Welander, A. Tillmar, Ø. Skare, T. Egeland, G. Holmlund, DNA microarray as a tool in establishing genetic relatedness - Current status and future prospects, Forensic Sci. Int. Genet. 6 (3) (2012) 322-329.

[9] D. Kling, T. Egeland, P. Mostad, Using object oriented Bayesian networks to model linkage, linkage disequilibrium and mutations between STR markers, PLoS ONE 7 (9) (2012) e43873.

[10] N. Sheehan, T. Egeland, Structured incorporation of prior information in relationship identification problems, Ann. Hum. Genet. 71 (2007) 501-518.

[11] D. Kling, A.O. Tillmar, T. Egeland, Familias 3—extensions and new functionality, Forensic Sci. Int. Genet. 13 (2014) 121-127.

[12] L. Poulsen, S. Friis, C. Hallenberg, B. Simonsen, N. Morling, Results of the 2011 relationship testing workshop of the English speaking working group, Forensic Sci. Int. Genet. Suppl. Ser. 3 (1) (2011) e512-e513.

[13] D. Balding, Weight-of-evidence for Forensic DNA Profiles, Wiley, New York, 2005.

[14] P. Gill, J. Buckleton. Biological Basis for DNA Evidence. J. Buckleton, C. Triggs, S. Walsh (Eds) *Forensic DNA Evidence Interpretation,* CRC Press, Florida, USA, 2005.

[15] W. Fung, Y. Hu, Statistical DNA Forensics: Theory, Methods and Computation, Wiley, England, 2008.

[16] I. Evett, B. Weir, Interpreting DNA Evidence, Sinauer, Sunderland, MA, 1998.

[17] D. Lucy, Introduction to Statistics for Forensic Scientists, Wiley, New York, 2006.

[18] C. Aitken, F. Taroni, J. Wiley, Statistics and the Evaluation of Evidence for Forensics Scientists, Wiley Online Library, New York, 2004.

[19] D. Primorac, M. Schanfield, Forensic DNA Applications: An Interdisciplinary Perspective, CRC Press, Boca Raton, FL, 2014.

[20] F. Taroni, C. Aitken, P. Garbolino, A. Biedermann, Bayesian Networks and Probabilistic Inference in Forensic Science, Wiley, England, 2006.

[21] R. Cowell, P. Dawid, S. Lauritzen, D. Spiegelhalter, Probabilistic Networks and Expert Systems: Exact Computational Methods for Bayesian Networks, Springer, Berlin, 2007.
[22] M. Hagmann, A paternity case for wine lovers, Science 285 (5433) (1999) 1470-1471.
[23] J.L. Gastwirth, J. Gastwirth, Statistical Science in the Courtroom, Springer, Berlin, 2000.
[24] L.H. Tribe, Trial by mathematics: precision and ritual in the legal process, Harv. Law Rev. (1971) 1329-1393.
[25] M.J. Saks, J.J. Koehler, The coming paradigm shift in forensic identification science, Science 309 (5736) (2005) 892-895.
[26] I.W. Evett, Interpretation: a personal odyssey, in: The Use of Statistics in Forensic Science, Ellis Horwood, Chichester, UK, 1991, 9-22.
[27] K. Slooten, R. Meester, Forensic identification: database likelihood ratios and familial DNA searching, arXiv preprint arXiv:1201.4261, 2012.
[28] G. Storvik, T. Egeland, The DNA database search controversy revisited: bridging the Bayesian-frequentist gap, Biometrics 63 (3) (2007) 922-925.
[29] P. Gill, Misleading DNA Evidence: Reasons for Miscarriages of Justice, Elsevier, London, 2014.
[30] A. Linacre, J.E. Templeton, Forensic DNA profiling: state of the art, Res. Rep. Forensic Med. Sci. 4 (2014).
[31] R. Elston, J. Stewart, A general model for the genetic analysis of pedigree data, Hum. Hered. 21 (6) (1971) 523-542.
[32] D. Gjertson, C. Brenner, M. Baur, A. Carracedo, F. Guidet, J. Luque, R. Lessig, W. Mayr, V. Pascali, M. Prinz, P. Schneider, N. Morling, ISFG: recommendations on biostatistics in paternity testing, Forensic Sci. Int. Genet. 1 (3) (2007) 223-231.
[33] E. Essen-Möller, Die Beweiskraft der Ähnlichkeit im Vaterschaftsnachweis. Theoretische Grundlagen, Mitteilungen der Anthropologische Gesellschaft (Wien) 68 (1938) 9-53.
[34] T. Egeland, B. Kulle, R. Andreassen, Essen-Möller and identification based on DNA, Chance 19 (2) (2006) 27-31.
[35] P. Gill, P.L. Ivanov, C. Kimpton, R. Piercy, N. Benson, G. Tully, I. Evett, E. Hagelberg, K. Sullivan, Identification of the remains of the Romanov family by DNA analysis, Nat. Genet. 6 (1994) 130-135.
[36] C. Schlötterer, D. Tautz, Slippage synthesis of simple sequence DNA, Nucleic Acids Res. 20 (2) (1992) 211-215.
[37] H. Ellegren, Microsatellites: simple sequences with complex evolution, Nat. Rev. Genet. 5 (2004) 435-445.
[38] A.P. Dawid, J. Mortera, V.L. Pascali, Non-fatherhood or mutation? A probabilistic approach to parental exclusion in paternity testing, Forensic Sci. Int. 124 (2001) 55-61.
[39] A.P. Dawid, J. Mortera, V.L. Pascali, D. van Boxel, Probabilistic expert systems for forensic inference from genetic markers, Scand. J. Stat. 29 (2002) 577-595.
[40] D. Balding, R. Nichols, A method for quantifying differentiation between populations at multi-allelic loci and its implications for investigating identity and paternity, Genetica 96 (1995) 3-12.
[41] D. Primorac, M. Schanfield, The Evaluation of Forensic DNA Evidence, National Academy Press, Washington, DC, 1996.
[42] P. Gill, L. Gusmao, H. Haned, W. Mayr, N. Morling, W. Parson, L. Prieto, M. Prinz, H. Schneider, P. Schneider, B. Weir, DNA commission of the International Society of

Forensic Genetics: recommendations on the evaluation of STR typing results that may include drop-out and/or drop-in using probabilistic methods, Forensic Sci. Int. Genet. 6 (6) (2012) 679-688.

[43] C. Brenner, B. Weir, Issues and strategies in the DNA identification of World Trade Center victims, Theor. Popul. Biol. 63 (3) (2003) 173-178.

[44] C. van Dongen, K. Slooten, M. Slagter, W. Burgers, W. Wiegerinck, Bonaparte: application of new software for missing persons program, Forensic Sci. Int. Genet. Suppl. Ser. 3 (1) (2011) e119-e120.

[45] K. Slooten, Validation of DNA-based identification software by computation of pedigree likelihood ratios, Forensic Sci. Int. Genet. 5 (4) (2011) 308-315.

[46] G. Dørum, D. Kling, C. Baeza-Richer, M. García-Magariños, S. Sæbø, S. Desmyter, T. Egeland, Models and implementation for relationship problems with dropout, Int. J. Leg. Med. 129 (3) (2015) 411-423.

[47] J. Curran, P. Gill, M. Bill, Interpretation of repeat measurement DNA evidence allowing for multiple contributors and population substructure, Forensic Sci. Int. 148 (1) (2005) 47-53.

[48] J. Buckleton, C. Triggs, Dealing with allelic dropout when reporting the evidential value in DNA relatedness analysis, Forensic Sci. Int. 160 (2) (2006) 134-139.

[49] T. Egeland, N. Pinto, M.D. Vigeland, A general approach to power calculation for relationship testing, Forensic Sci. Int. Genet. 9 (2014) 186-190.

[50] R. Szibor, M. Krawczak, S. Hering, J. Edelmann, E. Kuhlisch, D. Krause, Use of X-linked markers for forensic purposes, Int. J. Leg. Med. 117 (2) (2003) 67-74.

[51] E.A. Foster, M.A. Jobling, P.G. Taylor, P. Donnelly, P. De Knijff, R. Mieremet, T. Zerjal, C. Tyler-Smith, Jefferson fathered slave's last child, Nature 396 (6706) (1998) 27-28.

[52] A. Caliebe, A. Jochens, S. Willuweit, L. Roewer, M. Krawczak, No shortcut solution to the problem of Y-STR match probability calculation, Forensic Sci. Int. Genet. 15 (2015) 69-75.

[53] M.M. Andersen, P.S. Eriksen, N. Morling, The discrete Laplace exponential family and estimation of Y-STR haplotype frequencies, J. Theor. Biol. 329 (2013) 39-51.

[54] C.H. Brenner, Understanding Y haplotype matching probability, Forensic Sci. Int. Genet. 8 (1) (2014) 233-243.

[55] W. Bär, B. Brinkmann, B. Budowle, A. Carracedo, P. Gill, M. Holland, P.J. Lincoln, W. Mayr, N. Morling, B. Olaisen, et al., DNA Commission of the International Society for Forensic Genetics: guidelines for mitochondrial DNA typing, Int. J. Leg. Med. 113 (4) (2000) 193-196.

[56] R.N. Johnson, L. Wilson-Wilde, A. Linacre, Current and future directions of DNA in wildlife forensic science, Forensic Sci. Int. Genet. 10 (2014) 1-11.

[57] F.R. Bieber, C.H. Brenner, D. Lazer, Finding criminals through DNA of their relatives, Science 312 (5778) (2006) 1315-1316.

[58] J. Ge, R. Chakraborty, A. Eisenberg, B. Budowle, Comparisons of familial DNA database searching strategies, J. Forensic Sci. 56 (6) (2011) 1448-1456.

[59] S. Cowen, J. Thomson, A likelihood ratio approach to familial searching of large DNA databases, Forensic Sci. Int. Genet. Suppl. Ser. 1 (1) (2008) 643-645.

[60] J.M. Curran, J.S. Buckleton, Effectiveness of familial searches, Science & Justice 48 (4) (2008) 164-167.

[61] H.T. Greely, D.P. Riordan, N. Garrison, J.L. Mountain, Family ties: the use of DNA offender databases to catch offenders' kin, J. Law Med. Ethics 34 (2) (2006) 248-262.

[62] G. Miller, Forensics. Familial DNA testing scores a win in serial killer case, Science 329 (2010) 262.
[63] B. Olaisen, M. Stenersen, B. Mevåg, Identification by DNA analysis of the victims of the August 1996 Spitsbergen civil aircraft disaster, Nat. Genet. 15 (1997) 402-405.
[64] B. Leclair, R. Shaler, G.R. Carmody, K. Eliason, B.C. Hendrickson, T. Judkins, M.J. Norton, C. Sears, T. Scholl, Bioinformatics and human identification in mass fatality incidents: the World Trade Center disaster, J. Forensic Sci. 52 (4) (2007) 806-819.
[65] L.G. Biesecker, J.E. Bailey-Wilson, J. Ballantyne, H. Baum, F.R. Bieber, C. Brenner, B. Budowle, J.M. Butler, G. Carmody, P.M. Conneally, et al., EPIDEMIOLOGY: Enhanced: DNA identifications after the 9/11 World Trade Center attack, Science 310 (5751) (2005) 1122.
[66] C.H. Brenner, Some mathematical problems in the DNA identification of victims in the 2004 tsunami and similar mass fatalities, Forensic Sci. Int. 157 (2) (2006) 172-180.
[67] S. Donkervoort, S.M. Dolan, M. Beckwith, T.P. Northrup, A. Sozer, Enhancing accurate data collection in mass fatality kinship identifications: lessons learned from Hurricane Katrina, Forensic Sci. Int. Genet. 2 (4) (2008) 354-362.
[68] B. Leclair, C.J. Fregeau, K.L. Bowen, R.M. Fourney, Enhanced kinship analysis and STR-based DNA typing for human identification in mass fatality incidents: the Swissair flight 111 disaster, J. Forensic Sci. 49 (5) (2004) 939-953.
[69] B. Budowle, J. Ge, R. Chakraborty, H. Gill-King, Use of prior odds for missing persons identifications, Investig. Genet. 2 (2011) 1-6.
[70] L. Bradford, J. Heal, J. Anderson, N. Faragher, K. Duval, S. Lalonde, Disaster victim investigation recommendations from two simulated mass disaster scenarios utilized for user acceptance testing CODIS 6.0, Forensic Sci. Int. Genet. 5 (4) (2011) 291-296.
[71] P. Gill, Ø. Bleka, T. Egeland, Does an English appeal court ruling increase the risks of miscarriages of justice when complex DNA profiles are searched against the national DNA database? Forensic Sci. Int. Genet. 13 (2014) 167-175.
[72] K. Slooten, R. Meester, Probabilistic strategies for familial DNA searching, J. R. Stat. Soc. Ser. C (Appl. Stat.) 63 (3) (2014) 361-384.
[73] S.M. Suter, All in the family: privacy and DNA familial searching, Harv. J. Law Technol. 23 (2009) 309.
[74] T. Egeland, G. Dørum, M.D. Vigeland, N.A. Sheehan, Mixtures with relatives: a pedigree perspective, Forensic Sci. Int. Genet. 10 (2014) 49-54.
[75] N. Kaur, M. Bouzga, G. Dørum, T. Egeland, Relationship inference based on DNA mixtures. Int. J. Leg. Med., accepted for publication.
[76] Y.-K. Chung, Y.-Q. Hu, W.K. Fung, Evaluation of DNA mixtures from database search, Biometrics 66 (1) (2010) 233-238.
[77] Y.-K. Chung, W.K. Fung, Y.-Q. Hu, Familial database search on two-person mixture, Comput. Stat. Data Anal. 54 (8) (2010) 2046-2051.
[78] T. Egeland, I. Dalen, P. Mostad, Estimating the number of contributors to a DNA profile, Int. J. Legal Med. 117 (2003) 271-275.
[79] M. Kruijver, R. Meester, K. Slooten, Optimal strategies for familial searching, Forensic Sci. Int. Genet. 13 (2014) 90-103.
[80] L. Mayor, D. Balding, Discrimination of half-siblings when maternal genotypes are known, Forensic Sci. Int. 159 (2006) 141-147.
[81] T. Strachan, A. Read, Human Molecular Genetics, Garland Science, New York, 2010.
[82] J. Ott, Analysis of Human Linkage, third ed., The Johns Hopkins University Press, Baltimore, 1999.

[83] P. Gill, C. Phillips, C. McGovern, J.-A. Bright, J. Buckleton, An evaluation of potential allelic association between the STRs vWA and D12S391: implications in criminal casework and applications to short pedigrees, Forensic Sci. Int. Genet. 6 (4) (2012) 477-486.

[84] E.S. Lander, P. Green, Construction of multilocus genetic linkage maps in humans, Proc. Natl. Acad. Sci. 84 (8) (1987) 2363-2367.

[85] D. Kling, A. Tillmar, T. Egeland, P. Mostad, A general model for likelihood computations of genetic marker data accounting for linkage, linkage disequilibrium, and mutations, Int. J. Leg. Med. (2014) 1-12.

[86] G.R. Abecasis, J.E. Wigginton, Handling marker-marker linkage disequilibrium: pedigree analysis with clustered markers, Am. J. Hum. Genet. 77 (5) (2005) 754-767.

[87] A. Kurbasic, O. Hossjer, A general method for linkage disequilibrium correction for multipoint linkage and association, Genet. Epidemiol. 32 (7) (2008) 647-657.

[88] A.O. Tillmar, T. Egeland, B. Lindblom, G. Holmlund, P. Mostad, Using X-chromosomal markers in relationship testing: calculation of likelihood ratios taking both linkage and linkage disequilibrium into account, Forensic Sci. Int. Genet. 5 (5) (2011) 506-511.

[89] A.O. Tillmar, Population genetic analysis of 12 X-STRs in Swedish population, Forensic Sci. Int. Genet. 6 (2) (2012) e80-e81.

[90] G. Abecasis, S. Cherny, W. Cookson, L. Cardon, Merlin-rapid analysis of dense genetic maps using sparse gene flow trees, Nat. Genet. 30 (2002) 97-101.

[91] S. Friis, C. Hallenberg, B. Simonsen, N. Morling, Results of the 2013 relationship testing workshop of the English speaking working group, Forensic Sci. Int. Genet. Suppl. Ser. 4 (1) (2013) e282-e283.

[92] J.M. Curran, Introduction to Data Analysis with R for Forensic Scientists, CRC Press, Boca Raton, FL, 2011, ISBN 9781420088267.

[93] E. Thompson, Statistical inference from genetic data on pedigrees, in: NSF-CBMS Regional Conference Series in Probability and Statistics, JSTOR, 2000.

[94] N. Brümmer, Tutorial for Bayesian forensic likelihood ratio, arXiv preprint arXiv:1304.3589, 2013.

[95] P.J. Green, J. Mortera, Sensitivity of inferences in forensic genetics to assumptions about founding genes, Ann. Appl. Stat. (2009) 731-763.

[96] E. Thompson, Pedigree Analysis in Human Genetics, The Johns Hopkins University Press, Baltimore, 1986.

[97] K. Lange, Applications of the Dirichlet distribution to forensic match probabilities, Genetica 96 (1-2) (1995) 107-117.

[98] J. Butler, Forensic DNA Typing: Biology, Technology, and Genetics of STR Markers, Academic Press, London, 2005.

[99] Ø. Skare, N. Sheehan, T. Egeland, Identification of distant family relationships, Bioinformatics 25 (2009) 2376-2382.

[100] L. Gusmao, J.M. Butler, A. Carracedo, P. Gill, M. Kayser, W. Mayr, N. Morling, M. Prinz, L. Roewer, C. Tyler-Smith, et al., DNA Commission of the International Society of Forensic Genetics (ISFG): an update of the recommendations on the use of Y-STRs in forensic analysis, Forensic Sci. Int. 157 (2) (2006) 187-197.

[101] K.N. Ballantyne, V. Keerl, A. Wollstein, Y. Choi, S.B. Zuniga, A. Ralf, M. Vermeulen, P. de Knijff, M. Kayser, A new future of forensic Y-chromosome analysis: rapidly mutating Y-STRs for differentiating male relatives and paternal lineages, Forensic Sci. Int. Genet. 6 (2) (2012) 208-218.

[102] K. Slooten, F. Ricciardi, Estimation of mutation probabilities for autosomal STR markers, Forensic Sci. Int. Genet. 7 (3) (2013) 337-344.
[103] T. Egeland, N. Sheehan, On identification problems requiring linked autosomal markers, Forensic Sci. Int. Genet. 2 (2008) 219-225.
[104] D.J. Balding, M. Greenhalgh, R.A. Nichols, Population genetics of STR loci in Caucasians, Int. J. Leg. Med. 108 (6) (1996) 300-305.
[105] K.L. Ayres, D.J. Balding, Measuring departures from Hardy-Weinberg: a Markov chain Monte Carlo method for estimating the inbreeding coefficient, Heredity 80 (6) (1998) 769-777.
[106] T. Tvedebrink, P.S. Eriksen, H.S. Mogensen, N. Morling, Estimating the probability of allelic drop-out of STR alleles in forensic genetics, Forensic Sci. Int. Genet. 3 (4) (2009) 222-226.
[107] J.O. Berger, Statistical Decision Theory and Bayesian Analysis, Springer Science & Business Media, New York, 1985.
[108] A.O. Tillmar, P. Mostad, Choosing supplementary markers in forensic casework, Forensic Sci. Int. Genet. 13 (2014) 128-133.
[109] K. Slooten, T. Egeland, Exclusion probabilities and likelihood ratios with applications to kinship problems, Int. J. Leg. Med. 128 (3) (2014) 415-425.
[110] F. Riccardi, K. Slooten, Mutation models for DVI analysis, Forensic Sci. Int. Genetics 11 (2014), 88-95.
[111] M. Kruijver, Efficient computations with the likelihood ratio distribution, Forensic Sci. Int. Genet. 14 (2015) 116-124.
[112] K. Slooten, T. Egeland, Exclusion probabilities and likelihood ratios with applications to mixtures, Int. J. Leg. Med. First online: 10 July 2015.

Index

Note: Page numbers followed by *f* indicate figures and *t* indicate tables.

M

N

O

P

Q

R

S

T

U

W

X

Y

Z

Printed and bound by CPI Group (UK) Ltd, Croydon, CR0 4YY

11/06/2025

01899189-0009